Astronomers' Universe

Series Editor

Martin Beech, Campion College
The University of Regina
Regina, SK, Canada

The Astronomers' Universe series attracts scientifically curious readers with a passion for astronomy and its related fields. In this series, you will venture beyond the basics to gain a deeper understanding of the cosmos—all from the comfort of your chair.

Our books cover any and all topics related to the scientific study of the Universe and our place in it, exploring discoveries and theories in areas ranging from cosmology and astrophysics to planetary science and astrobiology.

This series bridges the gap between very basic popular science books and higher-level textbooks, providing rigorous, yet digestible forays for the intrepid lay reader. It goes beyond a beginner's level, introducing you to more complex concepts that will expand your knowledge of the cosmos. The books are written in a didactic and descriptive style, including basic mathematics where necessary.

Bernard Henin

Exploring the Ocean Worlds of Our Solar System

Second Edition

 Springer

Bernard Henin
Wiltshire, UK

ISSN 1614-659X   ISSN 2197-6651   (electronic)
Astronomers' Universe
ISBN 978-3-031-62955-6   ISBN 978-3-031-62953-2   (eBook)
https://doi.org/10.1007/978-3-031-62953-2

This Springer imprint is published by the registered company Springer Nature Switzerland AG
The registered company address is: Gewerbestrasse 11, 6330 Cham, Switzerland

If disposing of this product, please recycle the paper.

To my dearest companion, whose unwavering support has guided me through life's journey.

Preface

Approximately 1.2 billion kilometers away from our blue planet, frozen droplets of water orbit Saturn in unison with its majestic rings. These droplets are so abundant that they form a large ring around the planet. Hundreds of thousands of kilometers wide and 2000 km deep, this ring contains so many frozen water particles that Tethys and Dione, two small moons that happen to lie within it, have both developed a blue tint.

By analyzing this ring, the E ring, one of eleven other rings of Saturn (see Chap. 8), we have discovered that the droplets contain traces of salt (sodium chloride) and rocky matter (silicon dioxide), indicating that the body of water from which they originate must be warm, salty, and in direct contact with rocks very much like our seawater here on Earth. Science tells us that these conditions are favorable for life to develop and flourish, so it doesn't require a big stretch of the imagination to believe that, trapped inside these tiny seawater droplets, there might be microorganisms in the deep freeze. In other words, extraterrestrial life!

Scientists found the ocean from which the ring's frozen water particles originate, but this ocean is different from the ones we see here on Earth. It is a subsurface ocean that lies many kilometers beneath the surface of Enceladus, one of Saturn's icy moons. On this moon, mighty geysers regularly spout large amounts of salty water into space, thus, forming the E ring.

We now know that other worlds within our Solar System contain vast subsurface oceans. We call these planetary bodies 'Ocean Worlds', and they are one of the most exciting discoveries made in the last decades, thanks to the ongoing exploration of our Solar System.

It is truly extraordinary that we reside in an era where data obtained from robotic space missions enables us to engage in informed discussions regarding the potential existence of extraterrestrial life. In this book, we will embark on a journey through time, tracing the evolution of our understanding of these ocean worlds. We will then move through our Solar System and visit each ocean world, reviewing the evidence for their subsurface oceans and contemplate the intriguing concept of planetary habitability by analyzing the latest scientific data. Along the way, we'll ponder whether some of these oceans might harbor conditions conducive to life, and if so, what forms that life might take.

As such, this second edition retains the same structure as the first edition. It is divided into four parts, each focusing on a specific aspect of the ocean worlds. Part I, consisting of three chapters, covers basic concepts in planetary science and astrobiology to establish a good foundation upon which we can explore the subsequent chapters. Chap. 1 will reveal how the idea of ocean worlds was first introduced thanks to the remarkable journeys taken by NASA's Voyager spacecraft to the outer planets and their satellite systems in the last decades of the twentieth century, revolutionizing planetary science in the process. In Chap. 2, we will cover the origins of water in the universe as well as the processes behind its distribution throughout our Solar System. The possibility of life arising within subsurface oceans and the current approach that is being taken to find it will be described in Chap. 3. In so doing, we will review extremophiles on Earth and make a slight detour to the planet Mars, where the first ever interplanetary mission to detect alien life was undertaken in the 1970s.

With the essentials covered, our journey to the ocean worlds will start as we move into the second part of the book. There, we will explore in detail with the five confirmed ocean worlds of our Solar System, which are all icy moons of Saturn and Jupiter: Ganymede, Callisto, Europa, Titan, and Enceladus. Each one will have their own chapter, allowing us to explore their history as well as their physical and geochemical properties, and to ponder on the prospects of life within their subsurface oceans.

Part III will take us to three other icy moons and two dwarf planets where tantalizing clues suggest that a subsurface ocean, or smaller bodies of liquid water, lie under their icy crusts, but for which we still haven't found definitive proof. Within this part, Ceres, Dione, and Ariel will be covered in Chap. 9, while Triton and Pluto will be explored in Chap. 10. In the following chapter, we will explore numerous planetary objects that could theoretically have hosted a subsurface ocean in the past or might still do so in the present, but for which the limited observational data makes such cases debatable. This category includes, among others, icy moons such as Rhea, Titania, and Oberon as well as objects lying further out from Neptune such as Makemake, Eris, Sedna, and Gonggong.

Finally, the last part will review the proposed and current space missions designed to study the ocean worlds in the coming decades. In Chap. 12, we will look into the Juice mission by the European Space Agency and NASA's Europa Clipper mission, alongside other proposed missions awaiting selection. Given the life-detecting capabilities of future missions, we will end the chapter, and the book, speculating on the scientific and societal impact if we find evidence of extraterrestrial life within a subsurface ocean. Ultimately, looking for life forms in these remote and strange habitats is part of a bigger quest, the one for our cosmic origins.

In the appendix section, we will cover Mimas, a small moon of Saturn which had been previously put forward by some scientists as an ocean world candidate, only to be disproven. As such, this moon provides a cautionary tale on the drawbacks in interpreting from a limited set of data. In addition to Mimas, a brief overview of the hypothetical relic surface oceans of Mars and Venus will complete our investigation of past and present liquid water environments in our solar system.

Such a journey will also take us across the entire solar system where we will meet numerous planetary objects. From the now famous Comet 67P/Churyumov-Gerasimenko to Pluto's moon Charon. From Io, the most geologically active object in our Solar System to some of the remotest planetary objects known. We will venture far and wide, meeting in the process the robotic explorers that unveiled these worlds to us: Pioneer 10 and 11, Voyager 1 and 2, Galileo, Rosetta and Philae, Dawn, New Horizons, Cassini-Huygens, and more.

Throughout these chapters, we will cover the geological and geochemical processes involved in the alteration of planetary bodies such as the behavior of water in extreme conditions (Chap. 4) or external factors that alter a planetary surface exposed to space (Chap. 5). Further planetary processes and concepts will be distilled here and there throughout the book.

Key to the approach taken by this book is the fact that planetary science is a comparative science, where we gain much from comparing planetary objects with each other. As such, although it might be tempting to skip chapters and quickly jump to specific parts of the book (e.g., Europa), it is recommended to read in the order the chapters appear, as knowledge on the ocean worlds and the technology used to investigate them is reviewed progressively. Of course, in the case chapters are read individually, there will be pointers as to where a specific concept or technology has been covered elsewhere in more detail in the book.

In keeping with the comparative theme, every ocean world candidate mentioned in this book is presented in an overarching table, located after this preface, where comparisons on fundamental physical properties such as ratio or mass and the known or suggested characteristics of the subsurface oceans can be made between each candidate. The hope is that this table can be handy when one wants to quickly check the properties of a planetary object in relation to the ocean worlds concept, against what they might have just read or heard. Furthermore, a schematic diagram establishes where each ocean worlds candidate is located within the context of our solar system, making it easier for a novice to locate a given object.

Writing the second edition of this book is crucial due to the continuous evolution of planetary research. It allowed me to incorporate the latest findings, updates, and advancements, ensuring that the content remains relevant and accurate, thus upholding its value as a reliable resource for anyone interested in the field.

Sharing one's passion with others is one of life's greatest joys, and working on this second edition has provided me with another opportunity to do just that once again. I truly hope you will enjoy reading this book as much as I enjoyed researching and writing it. As with all my books, if anything written herein inspires you to learn more about space or science in general, then I've succeeded in my efforts.

Wiltshire, UK   Bernard Henin
August 2024

Acknowledgments

As can be expected with a project of such scope and depth, the insights and support of numerous people from have proved indispensable.

I would like to thank my publisher, Springer, for giving me this unique opportunity to share my passion for this fascinating topic and to inspire future generations of astronomers, space enthusiasts, and scientists. For the first edition of this book, John Watson, working for Springer in the U.K., proved instrumental in getting this project started and was a guiding hand throughout the course of the book proposal stage. I am incredibly grateful to Maury Solomon, my initial editor, who was the first, with John, to believe in this project from the outset and entrusted me with its writing.

For the second edition of this book, I express my deepest gratitude to Michael Maimone, my current editor, who was keen to get this new edition published and demonstrated endless patience towards me. Thank you for all your support.

In addition to my publisher, many people were involved in supporting me in my writing: Karen, Taryn, Pierre, Stephen, Chris, Steve, Maria, just to name a few.

Of course, I can't thank enough the scientists in the United States and Europe that took some of their precious time to contribute to this book via email exchanges, Skype, or phone interviews. The discussions I had with these leading scientists proved to be the highlight of this project. In alphabetical order, they are Dr. Penelope Boston, Dr. Charles Cockell, Dr. Amanda Hendrix, Dr. Luciano Iess, Dr. Jonathan Lunine, Dr. Chris McKay, Dr. William B. McKinnon, Dr. Marc Neveu, Dr. Olivier Witasse, and Dr. Steve Vance.

No acknowledgments would be complete without thanking my family, close friends, and everyone else who has supported me during the entirety of this project—you know who you are. I extend profound gratitude to my close family, to my dedicated wife who has graciously supported my enduring fascination with ocean worlds over the years, and to my daughter, who brings joy to my life each and every day.

Contents

About the Author

Bernard Henin Bernard Henin's fascination with planetary science ignited during his teenage years when he stumbled upon National Geographic images of Neptune captured by NASA's Voyager 2 spacecraft. Enthralled by the vastness of the giant blue planet, he felt a sense of exhilaration and liberation knowing that new worlds awaited exploration within his lifetime. Since that pivotal moment, Henin has avidly tracked humanity's ongoing journey to explore the mysteries of our Solar System. Henin is a professional science writer specialized in space and life sciences. He has authored two books for Springer Publishing on planetary science and space technology and serves as a contributing editor for the Sherwood Observatory in the United Kingdom, home to the second-largest telescope in the country that is freely accessible for public viewing. In addition, Bernard Henin has given numerous talks on space science to both astronomical societies and the public.Originally from Belgium, Henin has called Brussels, Houston, and Hong Kong home but have now planted his roots in the beautiful South West of England.

Contributors[1]

Penelope Boston NASA's Astrobiology Institute, Mountain View, CA, USA

Charles Cockell School of Physics and Astronomy at the University of Edinburgh, Edinburgh, UK

Amanda Hendrix Planetary Science Institute, Boulder, CO, USA

Luciano Iess Aerospace Engineering, Sapienza University of Rome, Rome, Italy

Jonathan Lunine Cornell University, Ithaca, NY, USA

Chris McKay NASA Ames Research Center, Mountain View, CA, USA

William B. McKinnon Department of Earth and Planetary Sciences, Washington University, St. Louis, MO, USA

Marc Neveu NASA Goddard Space Flight Center and University of Maryland, MD, USA

Olivier Witasse The European Space Research and Technology Centre, ESA, Noordwijk, The Netherlands

Steve Vance JPL, Pasadena, CA, USA

---

[1] I would like to express my most profound gratitude to the scientists listed who kindly found the time to talk to me and send me material. Without their contribution, making this book wouldn't have been possible.

Cross section of the five confirmed ocean worlds in our Solar System

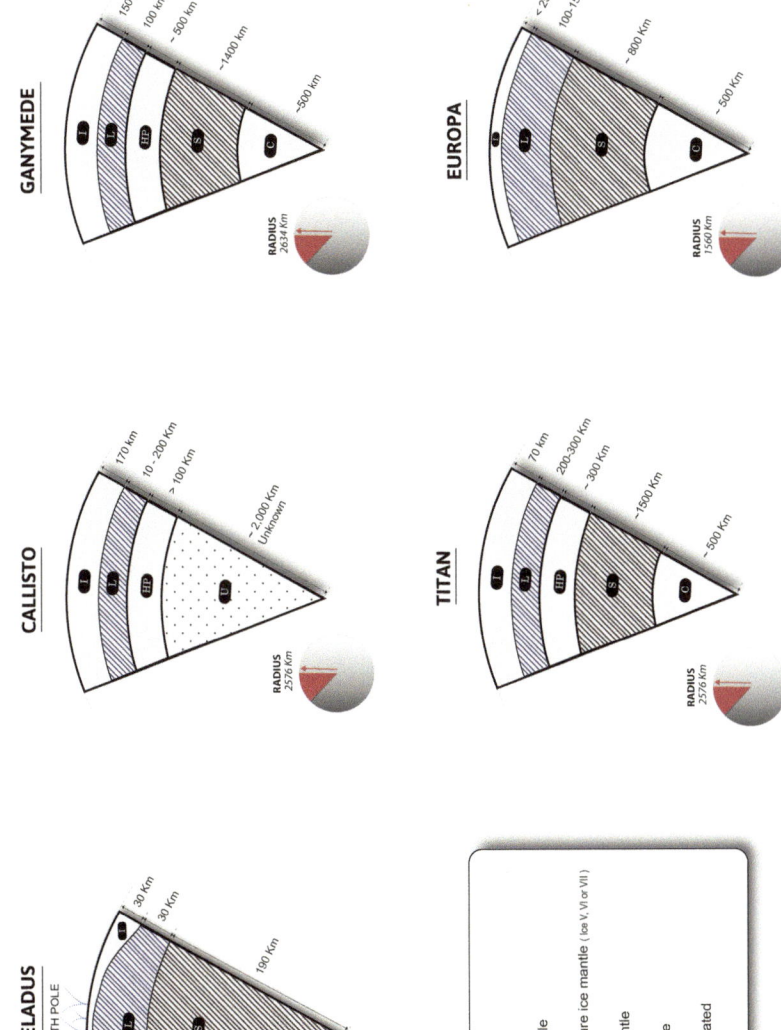

ENCELADUS
SOUTH POLE

RADIUS
252 Km

- 30 Km
- 30 Km
- 190 Km

CALLISTO

RADIUS
2576 Km

- 170 km
- 10 - 200 Km
- > 100 Km
- ~ 2,000 Km Unknown

GANYMEDE

RADIUS
2634 Km

- 150 km
- 100 km
- ~ 500 km
- ~1400 km
- ~500 km

TITAN

RADIUS
2576 Km

- 70 km
- 200-300 Km
- ~300 Km
- ~1500 Km
- ~ 500 Km

EUROPA

RADIUS
1560 Km

- < 25 Km
- 100-150 Km
- ~ 800 Km
- ~ 500 Km

I   Ice crust

L   Liquid mantle

HP   High pressure ice mantle (Ice V, VI or VII)

S   Silicate mantle

C   Metallic core

U   Undifferentiated

Diagrams are not to scale

Confirmed and potential ocean worlds in our Solar System

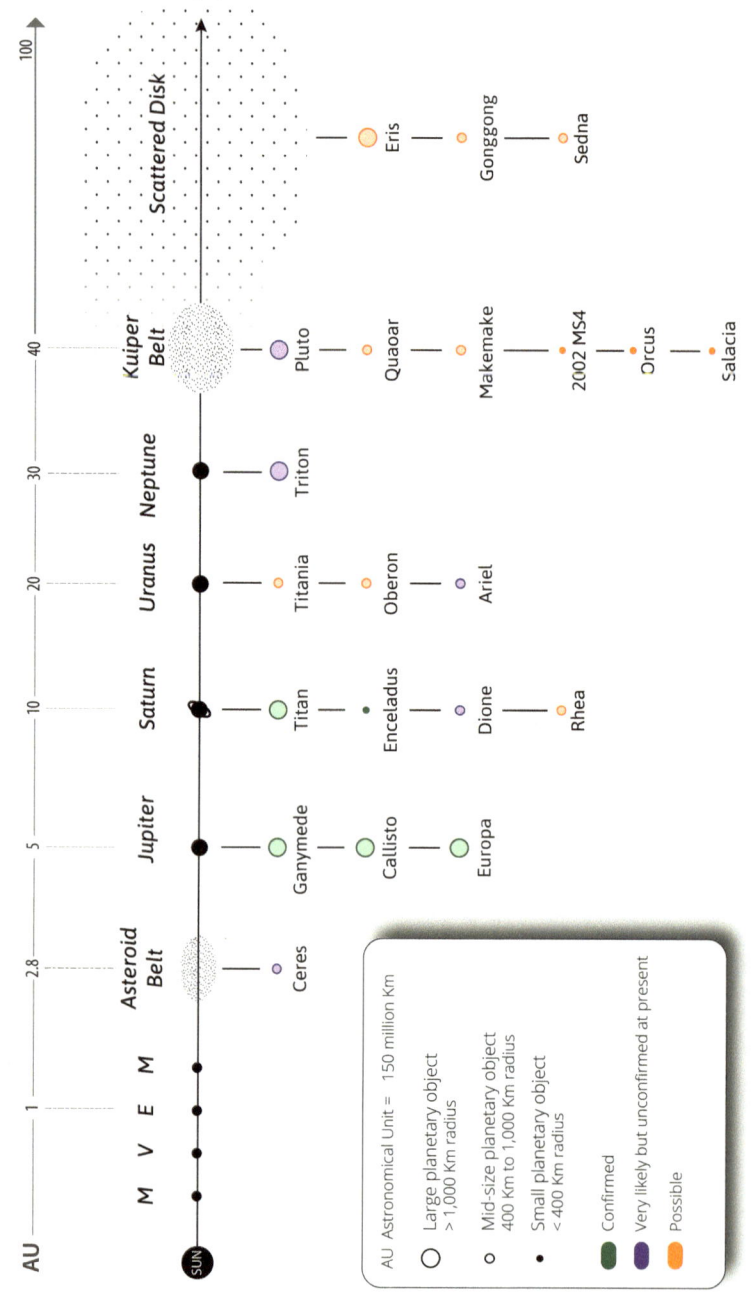

Diagram not to scale

Confirmed or Possible Ocean Worlds in our Solar System

Name of planetary object	Type of planetary object	Parent planet or Location	Distance from Sun (AU)	Mean Radius (km)	Mass $(10^{20}$kg)	Mean density (g/cm^3) Water = 1	Past or present subsurface ocean status	Lines of evidence of a subsurface ocean	Liquid water adjacent to rocky material	Future missions approved
Ceres	Asteroid, Dwarf planet	Asteroid belt	4	473	9	2.16	Very likely but unconfirmed	-	Most likely	-
Europa	Satellite	Jupiter	5	1,560	480	3.01	Confirmed	2	Yes	Juice, Europa Clipper
Callisto	Satellite	Jupiter	5	2,410	1,076	1.83	Confirmed	1	No	Juice
Ganymede	Satellite	Jupiter	5	2,634	1,482	1.94	Confirmed	1	No	Juice
Enceladus	Satellite	Saturn	10	252	1	1.61	Confirmed	+2	Yes	-
Titan	Satellite	Saturn	10	2,575	1,346	1.88	Confirmed	2	No	Dragonfly
Rhea	Satellite	Saturn	10	764	23	1.23	Possible	-	-	-
Dione	Satellite	Saturn	10	560	11	1.47	Very likely but unconfirmed	-	Most likely	-
Ariel	Satellite	Uranus	20	579	14	1.59	Very likely but unconfirmed	-	-	-
Titania	Satellite	Uranus	20	788	35	1.71	Possible	-	-	-
Oberon	Satellite	Uranus	20	761	30	1.63	Possible	-	-	-
Triton	Satellite	Neptune	30	1,353	214	2.06	Very likely but unconfirmed	-	Most likely	-
Makemake	KBO, Dwarf planet	Kuiper Belt	40	~720	44	1.4 - 3.2	Possible	-	-	-
2002 MS4	KBO	Kuiper Belt	40	~467	-	-	Possible	-	-	-
Quaoar	KBO	Kuiper Belt	40	~537	~14	~2.2	Possible	-	-	-
Salacia	KBO	Kuiper Belt	40	~425	~4.4	~1.29	Possible	-	-	-
Orcus	KBO	Kuiper Belt	40	~460	~6.4	~1.5	Possible	-	-	-
Pluto	KBO, Dwarf planet	Kuiper Belt	40	1,188	130	1.85	Very likely but unconfirmed	-	Most likely	-
Eris	SDO, Dwarf planet	Scattered disk	30-100	1,163	166	2.52	Possible	-	-	-
Sedna	SDO	Scattered disk	30-100	~498	-	-	Possible	-	-	-
Gonggong	SDO, Dwarf planet	Scattered disk	30-100	~751	-	-	Possible	-	-	-

AU : Astronomical Unit / KBO : Kuiper Belt Objects / SDO : Scattered Disk Objects

Part I
The Origin of Water and Life

Equipped with his five senses, man explores the universe around him and calls the adventure Science. (Edwin Powell Hubble)

In Part I, we review the revolution that occurred in planetary science when the Voyager space probes visited the outer planets and their satellite systems, bringing back the first hints of the ocean worlds in our solar system. The second chapter deals with the origin of water in space and how it was distributed among the planetary objects orbiting our Sun, while the third chapter deals with the possibility of extraterrestrial life and our attempts to find it.

Chapter 1
The Voyagers' Tale

Golden Amazons of Venus

The night sky has always been a source of fascination for humankind. For centuries, storytellers have turned to it to create fantastical visions, yet, as we further pushed the boundaries of scientific exploration, new worlds emerged.

When astronomers first pointed their telescopes at the Moon in the seventeenth century, they assumed that they were looking at a world awash with liquid water and assigned the watery names to its surface features: Maria (singular mare, Latin for "sea"), Oceanus (singular oceanus, Latin for "ocean"), Lacus (singular lacus, Latin for "lake"), Sinus (singular sinus, Latin for "bay") and Paludes (singular palus, Latin for "marsh"). Yet, the Eagle that landed in the 'Sea of Tranquility' did so on struts rather than floats.

Similarly, the discovery of an atmosphere around Venus in 1761 led to speculation that hidden beneath the thick Venusian cloud cover, was a lush and humid world. Soon, Venus as a 'water world' captured the imaginations of astronomers and science fiction writers alike. A quick browse through some of the early science fiction novels reveals titles such as" Oceans of Venus" by Isaac Asimov, "Swamp Girl of Venus" by H. H. Harmon, and the classic "Golden Amazons of Venus" from J. M. Reynolds. The last two titles are from the so-called pulp era of science fiction in the 1930s and 1940s, where scientific observations were often side-lined by extravagant fantastical stories, now referred to as planetary romance.

The age of Venusian blondes waiting to be rescued by virile Earthlings ended abruptly in 1962 when NASA's Mariner 2 spacecraft completed the first-ever flyby of the planet (or any planet for that matter). Recording atmospheric temperatures of 500 °C, there was no escaping the fact that the surface of Venus is hot enough to melt lead and that, sadly, there are no seas on Venus of liquid water and no Venusians.

A similar story followed with Mars, the red planet, which has long been a source of intrigue. Mars was first observed through a telescope in 1610 by Galileo Galilei,

© The Author(s), under exclusive license to Springer Nature Switzerland AG 2024
B. Henin, *Exploring the Ocean Worlds of Our Solar System*, Astronomers' Universe,
https://doi.org/10.1007/978-3-031-62953-2_1

the father of observational astronomy. Unfortunately, his telescope wasn't powerful enough to reveal the planet's distinct surface features and we had to wait until 1659 when Christian Huygens, a Dutch astronomer, drew a rudimentary map of Mars' surface using a telescope he built himself. The map showed darkened areas, convincing Huygens that these were signs of vegetation, which led him to publish his belief in extraterrestrial life in his influential book Cosmotheoros. He was also the first man to see the white Martian south polar cap but didn't know what to make of it. More than a century passed before it was correctly identified as water-ice by Sir William Herschel, a German-born British astronomer, who also postulated that the dark areas on Mars were oceans, similar to the lunar mares. Herschel's work on Mars and the realization that it seemed to share many similarities to our own further boosted the credibility to the idea that life was also present on the red planet. He even speculated that Martian inhabitants "probably enjoy a situation similar to our own."

The belief that water was flowing on Mars reached its height in the early twentieth century where a sloppy translation from Italian to English led to the belief that irrigation canals had been built all across the planet. The excitement died down over the course of the century as technological advances gave astronomers the ability to see the planet in further detail. The idea of a Martian civilisation struggling to cope with a drying world was finally laid to rest in 1972 when NASA's Mariner 9 spacecraft returned images of a lifeless, desolate planet.

As more spacecraft ventured outwards, our solar system was shown to be inhospitable and barren. Gone were the wondrous Selenites, the attractive Venusians and the hardy Martians. As far as we looked, our blue oasis was the only place that could support life, and science fiction, one of the most imaginative and thought-provoking genres, had reached an impasse. As a result, swashbuckling spacemen moved on to the more promising lands outside of our solar system with the help of warp engines and tricorders. Earth's neighboring planets and their moons were shunned.

This would not last. Our understanding of the solar system changed once again and, as the title of this book gives away, evidence of liquid water was found, albeit in less obvious places: within the moons of the outer planets. Vast oceans of flowing water lie waiting to be explored.

The discovery of these hidden oceans started as the two Voyager spacecraft, conceived when our solar system was thought to be barren, embarked on long journeys that would perform flybys of the Jovian moons as their first stage. These close encounters would create a revolution.

Actually, despite their relatively small sizes when compared to the outer planets, the satellites of Jupiter had already been game changers in the past and played a remarkable role in the history of astronomy, science and our understanding of humanity's place within the universe, no less.

Described by Galileo Galilei in January 1610 as "three fixed stars, totally invisible by their smallness," they were found to be very close to Jupiter and seemingly positioned in a straight line across it. This configuration, and the fact that the 'stars' disappeared behind Jupiter only to reappear once again later, led the Galileo to deduce that these were, in fact, moons. This straightforward yet significant discovery made the Italian astronomer the first person to see and understand that objects were orbiting other planets, and with this new paradigm, the unravelling of the Tychonic system[1] commenced.

Galileo published his findings in March 1610 in a book titled "Sidereus Nuncius" or "The Starry Messenger." Not an imprudent man, he originally named these four moons as the Medicea Sidera ("the Medician stars"), honouring his patron and the other Medici brothers: Cosimo, Francesco, Carlo, and Lorenzo. Thankfully, or regrettably, these names were replaced in 1614 by Simon Marius, a German astronomer, who named the moons after Zeus's lovers in Greek mythology: Io, Europa, Ganymede, and Callisto.

And so, fast forward to 1979, almost 400 years after their discovery, Jupiter's moons would once again change our understanding of the worlds surrounding us. This time though, it wasn't done with Earth-based observations using powerful telescopes, but with robot probes sent on a multi-year journey to image the outer planets and their moons using onboard cameras.

As such, only 20 years after the Soviets sent the very first artificial object into space, the United States launched not one but two spacecraft: Voyager 1 and Voyager 2. Taking advantage of a favorable alignment of the outer planets (next occurring in the year 2153), these new emissaries embarked on a grand tour, visiting not only Jupiter but Saturn, Uranus, and Neptune, too.

Prior to the Voyagers' grand tour, the only moon we knew relatively well was our own, whose official name is "Luna." Although magnificent to look at, our Moon is geologically inactive. And while most members of the public assumed that the other moons of our solar system would be similar to Luna, scientists knew otherwise, as they had already been analyzing the many moons' reflected light, known as spectra, through Earth-based observations.

These observations revealed not only that specific moons had icy surfaces, contrary to Luna, but that they also displayed albedo and color variations as they rotated suggesting diverse geological terrains. Because of this, scientists were eager to know more and lobbied heavily for robotic probes to study the moons of the outer solar system. Yet, with very little to go by, planning these future missions were mainly based on guesses. Some were accurate, others less so.

For example, when the Voyager missions were being prepared, Jupiter's moon Europa (see Chap. 6) was thought to be of little importance compared to the other Galilean satellites, as it was the smallest of the four. Io was by far the most intriguing object with its colorful surface features faintly observed from ground telescopes. Ganymede and Callisto were so big that their size alone was a key attraction; let us not forget that Ganymede is bigger than Mercury and almost as big as Mars. So, when it came to planning the routes of the Voyagers through the Jovian system, Europa was at the bottom of the list, not warranting a close flyby.

As we now know, scientists were in for a big surprise. When Voyager 1 first reached the Jovian system in 1979 and flew past Europa, at the intended distance of

---

[1] From the ancient Ptolemaic system suggesting that Earth was at the center of the universe.

2 million kilometers, the low-resolution images returned by the spacecraft bewildered planetary scientists (Fig. 1.1).

The images returned a bright moon crisscrossed by mysterious intersecting linear features that had not been predicted. Also, most scientists expected that small celestial bodies such as Europa would show a heavily cratered surface similar to Luna. The thinking went that these small bodies lacked sufficient heat to support active geology which had the potential to reshape surfaces and erode or erase craters. Yet, impact craters were few on Europa. How could that be? Furthermore, dark patches were detected on the moon's surface, however, few scientists had an idea of what these might be. This was all very puzzling.

Through its density, derived from the mass and volume, and light spectrum, Europa was thought to be a rocky moon with a relatively thin layer of water at its surface, forming an icy shell. Could the linear features observed on the surface be deep cracks within the ice crust caused by unknown tectonic processes? Could it be that Europa was still geologically active? Such a notion as thought provoking.

Fortunately, Voyager 2 would make a closer flyby 4 months later and return high-resolution images from the surface (Fig. 1.2).

The images returned by Voyager 2 allowed scientists to study the moon's surface in far greater detail and impact craters were counted more precisely, revealing that Europa had far fewer craters than Callisto and Ganymede. The evidence was unambiguous: contrary to most expectations, Europa's icy crust was young. Very young. Maybe less than a hundred million years old; a relatively short span of time in planetary science.

Fig. 1.1 Europa, the icy moon of Jupiter, viewed by Voyager 1 on March 4, 1979. This shot was the best resolution obtained by the spacecraft. We can see bright areas contrasting with dark patches, crisscrossed by long linear structures. (Image courtesy of NASA/JPL)

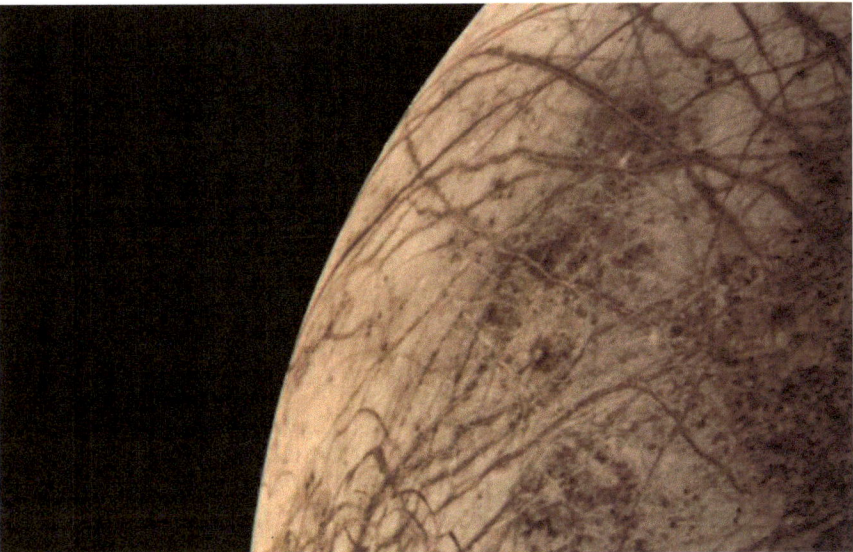

Fig. 1.2 Taken by Voyager 2 on July 9, 1979. A closer look at Europa revealed few impact craters and a complicated, fractured crust. The lack of any mountains or craters is consistent with an active ice crust. (Image courtesy of NASA/JPL)

In addition, the surface was much smoother than expected, displaying very little height variation with no mountains or tall hills. This can only be explained if a surface is too elastic to retain tall features such as crater rims or cryovolcanoes (volcanoes formed by volatiles such as water, methane, or ammonium). Somehow, Europa's icy crust wasn't as hard solid as what might have been expected from an object lying far away from the Sun. The images returned by Voyager 2 were inexorable. Europa, the relatively small icy moon of Jupiter, was a geologically active planetary body capable of resurfacing itself.

That such a small planetary body could retain enough heat to present such active surface features puzzled many scientists. Yet, one hypothesis proposed prior to the Voyagers' flybys soon gained the attention of the scientific community. It suggested that, depending on a moon's orbit, enough heat could be generated deep within the interior to melt the present rocks or ices, depending on the moon's composition.

In the case of Europa, the ice mantle could theoretically be warmed up, thus, creating vast amounts of liquid water upon which an icy crust, exposed to coldness of space, would rest—in other words, a subsurface ocean would be formed. In addition, tidal heating suggested that heat exchanges between the subsurface ocean and the icy crust above it could deform and stress the surface, thus creating cracks and erasing craters. Could this new theory be the cause of Europa's unusual surface features? The scientific community was abuzz.

A New Type of Energy Source

To understand tidal heating though, we must go back to when the Voyagers made close flybys of Io, one of Europa's neighboring moon and the closest to orbit Jupiter. As stated above, Io had been listed as a planetary body of great interest for the Voyager missions; this was due to images taken 5 years earlier by another, simpler, American spacecraft called Pioneer 11, which hinted at a bright yellow and orange colored, yet undetermined, surface. This, and other intriguing facts about Io had piqued scientists' interest, and the trajectories of both Voyagers were conceived in such a way that close flybys would take place.

So, when Voyager 1's high-resolution images of Io finally came back, see Fig. 1.3, scientists were stunned as the moon was found to be an active world; not of ice, but of fire! Io's surface was found to be peppered with volcanoes upon which eruption plumes kilometres high and lava flows, stained yellow and red by oxides of sulfur, emerged. Tall mountains and calderas were observed as well. Remarkably, not a single impact crater was detected on the images, implying continual resurfacing due to the ever-present volcanic activity. Io became overnight a volcanologist's dream world.

Following Io's flyby by Voyager 1, and the discovery of the enigmatic source of energy, scientists were eager to find an explanation. Luckily, it came from a

Fig. 1.3 A fiery Io captured by Voyager 1 on March 4, 1979, the same day that the spacecraft took its best resolution image of Europa. The distance to Io is about 490,000 km. A volcanic explosion can be seen in the upper left ejecting solid material to an altitude of 160 km. (Image courtesy of NASA/JPL)

scientific paper published in the prestigious journal Science just a few days before the Voyagers' arrival in the Jovian system. The paper, written by Stanton Peale and his colleagues, suggested that heat might be significant in Io's interior as the moon orbits its parent planet, Jupiter, in a highly elliptical orbit. Note that elliptical orbits are measured by their eccentricity; the greater the eccentricity, the more elliptical the orbit will be and vice versa. Eccentric orbits produce variations in the gravitational pull from the parent planet. Peale's paper suggested that such variations induce tidal heating.

With that in mind, what goes on inside Io can be easily demonstrated by using a simple metal wire like a paperclip. If you happen to have one to hand, flex one part of the wire backwards and forwards. You will notice that it doesn't take long for heat to appear on the bendy part. The explanation is simple: the kinetic energy applied by the movement of your fingers was transformed into thermal energy through internal friction. A similar process makes squash balls quite warm after a match.

The reason why some of the Galilean moons, such as Io, are subjected to a sizeable amount of tidal heating is in part due a phenomenon known as orbital resonance. This locks each moon into a specific orbital ratio around their parent planet. As such, for every two orbits that Io takes around Jupiter, Europa takes precisely one. But there's more to this. Due to the action of orbital mechanics, Io and Europa always come closest to each other at the same location within their orbits, just like clockworks. Thus, Io and Europa pull at each other, resulting in a more elliptical orbit instead of a circular one. (Similarly, for every two orbits that Europa takes around Jupiter, Ganymede makes precisely one. This 4-2-1 sequence dictates the orbital eccentricity of the three inner Jovian moons as we shall see in subsequent chapters.)

With Io's orbit characterised by eccentricity, the moon will feel Jupiter's gravitational pull differently along its orbit. This is referred to as tidal forces and is similar to the gravitational effect our Moon has on the seas and oceans of Earth. On Io, the tidal forces will be strongest during the moon's closest approach in orbit (periapsis) than during its furthest point (apoapsis). As the moon moves from periapsis to apoapsis and back, the tidal forces pull its interior at varying intensities and the rock and iron mantles repeatedly distort and buckle, creating friction. Thus, heat is generated.

Many factors determine how tidal forces will influence a planetary body, such as the size of the moon in relation to its parent planet as well as the distance of the moon's orbit. As importantly though, the moon's composition and rigidity will determine how it responds to the gravitational sways; if it is rocky like our Moon, it will distort far less than if it is made entirely of ice which is far more malleable. We can measure the rigidity of a planetary body and its ability to change in response to a tidal potential thanks to a property called the Love number introduced in the early twentieth century by the famous British mathematician Augustus Edward Hough Love. Zero is for a rigid body. In the case of Io, the love number is not a constant value, but varies depending on the location. The latest available data suggests that Io's love number ranges from around 0.02 to 0.1.

Now that Io's power source was well understood, it was generally agreed upon that Europa was also subjected to tidal heating due to its known resonance with Ganymede and Io. Could the heat generated by the tidal forces be truly capable of melting parts of Europa's thin icy crust and—gasp—create a subsurface ocean? No one could tell for sure, and this was undoubtedly the central thesis proposed to explain the moon's deformed surface. Future investigations would be required to test this idea.

After Io and Europa though, planetary scientists turned their attention to the planet-size moons Ganymede and Callisto. Ganymede's surface didn't have Europa's pizzazz, but it did show two distinct terrains: one dark and cratered and therefore old, and the other grooved, with fewer craters implying recent geological or tectonic activity. Could the relatively younger terrain be a result of tidal heating? And, if so, was heat still being generated within its interior, like Io and Europa, and would this activity be sufficient to create and maintain a subsurface body of water? Unfortunately, none of these questions could be answered with any confidence with the data returned by the Voyagers' flybys. We would have to wait for future missions to start providing some answers (Chap. 4 reviews Ganymede in more detail) (Fig. 1.4).

Fig. 1.4 This picture of Ganymede was taken on March 5, 1979, by Voyager 1 at a distance of 272,000 km. The bright areas contain grooves and ridges indicating geological activity, while many older impact craters have been eroded over time. (Image courtesy of NASA/JPL)

Callisto, the last of the Galilean moons, displayed very little eccentricity in its orbit due to a weaker orbital resonance pattern. For every three orbits Callisto takes around Jupiter, the neighboring Ganymede takes seven. This 'imperfect' orbital pattern, and the fact that Callisto's orbit lies much further away from Jupiter, implied that the giant moon wouldn't experience much tidal heating. The images returned from the Voyagers showed Callisto's surface to be one of the most heavily cratered surfaces in the solar system, with no signs of past or present geological activity. Compare this to Io, the most geologically active body in our solar system, and you find a scale within the Galilean moons: the further away they are from Jupiter, the less energy they gain through their orbits. Despite this, could Callisto harbor a subsurface body of water as well? Again, we would have to wait for future missions to answer this (see Chap. 5 for further details on Callisto) (Fig. 1.5).

Fig. 1.5 Callisto as seen by Voyager 2 on July 7, 1979, at a distance of 1 million kilometers. Variations of surface materials can be seen in UV. The moon of Jupiter hosts the most densely cratered surface in our solar system. (Image courtesy of NASA/JPL)

The Moons of Saturn

What the Voyagers had unveiled as they flew past Jupiter and its satellites transformed planetary science overnight; a new energy source capable of heating up the small icy moons of our solar system had been discovered. Scientists could once again contemplate the existence of liquid water far from Earth. As the Voyagers rushed towards their next destination, people started to wonder what discoveries they would make there.

The Saturnian system was a rich target. It has a vast weather system, many times bigger than Earth's. It has a grandiose set of rings requiring detailed observations. It has Titan, the only moon in our solar system known to support a thick atmosphere. And it has strikingly bright and tiny moons packed with water-ice and rocks such as Enceladus (see Chap. 8 for further details).

Enceladus had already raised scientists' eyebrows 14 years earlier. Images of the E ring with ground-based observations had managed to image Saturn's E ring, revealing that its densest part was centered on the tiny moon's orbit. Whatever was happening on Enceladus, it was clear to all that it was the source of the particles forming the E ring. Planetary scientists awaited eagerly as the Voyagers picked up speed at Jupiter and carried on their interplanetary voyage towards Saturn (Fig. 1.6).

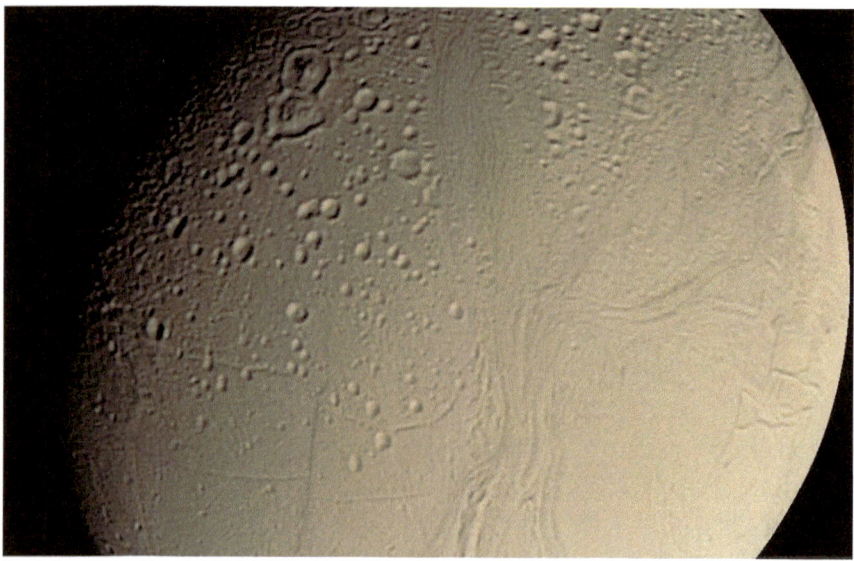

Fig. 1.6 This color image of Enceladus, one of Saturn's icy moons, is a mosaic of Voyager 2 images taken in August 1981. The moon reflects 90% of incident sunlight, making it the most reflective object in the solar system. (Image courtesy of NASA/JPL/USGS)

The Voyagers were not the first to reach the ringed planet though. NASA's Pioneer 11 had already conducted Saturn's first ever flyby in 1979, yet the low-resolution images it sent back weren't detailed enough to characterize the moons' surfaces, so little insight on Enceladus had been gained. Luckily, scientists didn't have to wait long. Voyager 1 would arrive in the Saturnian system in 1980, and Voyager 2 would be following 9 months later.

Without doubt, one of the many highlights of the Voyager mission were the close-up images of Saturn's icy moons, and especially of Enceladus. The Voyagers' high-resolution images revealed its surface to be unusually smooth, with very little cratering apparent. Was Enceladus also being subjected to tidal stresses like Europa?

The scientific community was excited by such possibility and the moon quickly became one of the most fascinating planetary bodies in our solar system. But there was just one problem with this explanation: Enceladus, with a diameter of 504 km, is six times smaller than Europa and has a relatively low orbital eccentricity (0.0047). That's half of what Europa experiences. When scientists considered these factors, their calculations showed that tidal heating was insufficient to explain the present activity observed on the moon. Although various proposals were put forth to explain the discrepancy, no consensus could be reached amongst the scientific community and the source of Enceladus' heat, and its smooth surface remained a mystery. It would be so for many years.

Other Saturnian moons also proved interesting. Mimas, the innermost of Saturn's major moons, is less than 198 km in mean radius, making it the smallest spherical body in our solar system. In fact, it is so tiny that it can barely maintain its shape although this wasn't the only characteristic that made Mimas special. With an orbital eccentricity four times that of Enceladus and a much closer orbit to Saturn, theoretical models predicted that ice-packed Mimas should experience more tidal heating than Enceladus.

Yet, when the mission scientists got their first glimpses of Mimas, they were surprised to discover one of the most densely cratered surfaces in the solar system. The scientific community was now faced with the opposite problem they had encountered with Enceladus: instead of looking at a geologically active body pumped up by tidal heating, they were looking at a frozen moon whose surface had remained unchanged for billions of years. Something was amiss.

This contradiction didn't prevent some scientists from suggesting that liquid water could still exist deep within the Mimas' interior, although most considered such possibility remote. It was clear, though, that these were early days, and that additional scientific data would be required to resolve this paradox. Planetary scientists would have to wait 20 years to learn more. (More details on Mimas can be found in the appendices.) (Fig. 1.7)

Another intriguing icy moon revealed by the Voyagers was Dione. Sort of a bigger sister to Enceladus, Dione exhibited a less active history as its surface shows heavily cratered regions, although moderate or lightly cratered plains can be found as well. In addition, the images returned by the Voyagers revealed mysterious wispy material composed of bright, narrow lines, leading some scientists to suggest that these could be the result of fresh ice seeping from the interior of the moon.

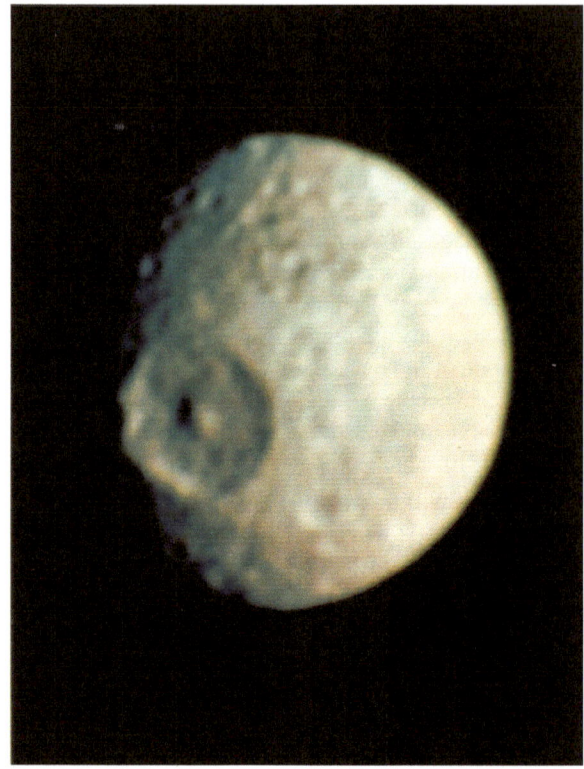

Fig. 1.7 Taken from half a million kilometers away, this is one of the first pictures of Saturn's moon Mimas, made by Voyager 1 during its flyby on November 12, 1980. The massive crater, approximately a 100 km wide and therefore about one-quarter of the satellite's diameter, is named after the eighteenth-century astronomer William Herschel, who discovered Mimas in 1789. (Image courtesy of NASA/JPL)

Dione is in orbital resonance with Enceladus, completing two orbits of Saturn for every single orbit completed by Enceladus, giving it an orbital eccentricity of 0.0022. That's half of Enceladus' eccentricity. While some signs of early geological activity did suggest the existence of liquid water under Dione's icy crust in the past, the possibility that the energy generated by tidal heating could have prevented it from freezing up seemed faint (Fig. 1.8).

Finally, in March 1979, one of the most awaited events of the entire mission took place: Voyager 1's flyby of Titan. This giant moon hidden beneath a shroud of orange atmosphere was full of possibilities. Larger than the planet Mercury, and laced with organic gases, Titan was thought to have a liquid cycle of methane (lakes, rain, and gas). It was such a unique body in our solar system that scientists had decided early in the planning of the Voyager missions that Voyager 1 would be making a very close flyby of the giant moon.

Unfortunately, the constraints of orbital mechanics meant that such a flyby would slingshot the spacecraft onto a trajectory outside of the ecliptic plane, ruling out any further visits to the other outer planets as no other routes allowed a close pass of Titan while preserving a Uranus flyby option. Once Voyager 1 would have performed its Titan flyby, Voyager 2 would be the only spacecraft that could continue its exploratory mission into the outer solar system. There would be no backup plan.

Fig. 1.8 Dione viewed by
Voyager 1 from a distance
of 160,000 km on
November 12, 1980. The
wispy material can be seen
on the edges of the small
moon. (Image courtesy of
NASA/JPL)

Titan proved to be such an important target that despite the risks, the mission planners decided to go ahead with the flyby. In fact, the Voyager mission was planned in such a way that if Voyager 1 were to fail in completing its objectives at Titan, Voyager 2 would be reprogrammed to make the flyby instead, prematurely ending the tour as neither Uranus nor Neptune would be visited. It is telling that Titan was thought to be a more important from a scientific point of view than Uranus and Neptune with their entire system of satellites.

A year prior to Voyager 1's flyby of Titan, Pioneer 11 had passed within 355,600 km of the moon. Unsurprisingly, the spacecraft had returned low-resolution pictures with its limited imaging capabilities, presenting it as a featureless orb. Little could be deduced at the time. So, when, in November 1980, Voyager 1 passed at only 3915 km—the closest approach to a moon or planet by the Voyager mission—scientists were hoping to learn more from Titan's mysterious surface. Unfortunately, the Voyager 1 pictures disappointed as they showed a thick, impenetrable atmosphere with no breaks within the clouds. Titan's surface would remain enigmatic for now.

Voyager 1 nevertheless returned promising scientific data. Its spectrometers analysed the moon's atmosphere and detected a variety of organic compounds in the atmosphere. It also discovered that at 0.0288, Titan experienced the strongest orbital eccentricity of all the moons of Saturn and Jupiter. With a density between that of solid rock and water, the moon's interior was thought to be composed of a thick mantle of ice sitting atop a rocky core. Would this ice layer be subjected to the effects of tidal heating? The scientists didn't know. By then, new theoretical models of the moon's interior showed that, under the right conditions, a layer of water ice could have melted. Titan had joined the list of planetary bodies that might have a subsurface body of liquid water, thus, there was a potential for it to host two entirely different liquid environments: liquid methane on its surface (due to the environmental conditions expected to be present there) and liquid water within its interior. Sadly, it would be 24 years before another spacecraft would finally begin to unveil Titan's liquid promises.

Beyond Saturn

And so, as the Voyagers 1 and 2 left the Saturnian system, the latter on its way to Uranus and Neptune, the former flying straight out of the solar system on a trajectory perpendicular to the ecliptic plane, planetary science had been transformed in just a few short years. Instead of a dry, inert, and unexciting set of moons lying far away from the Sun, the Voyagers had found an assortment of fascinating and geologically active worlds around Jupiter and Saturn that had the potential to host vast subsurface oceans. The solar system was sloshy.

Table 1.1 This table represents the exploration of the outer planets since the first voyage of the Pioneer probes in 1973. NASA holds the title of being the only space

Table 1.1 Outer planets exploration chart.

		Spacecraft	Jupiter	Saturn	Uranus	Neptune	Pluto
NASA	Launched	Pioneer 10	1973—flyby				
		Pioneer 11	1974—flyby	1979—flyby			
		Voyager 1	1979—flyby	1980—flyby			
		Voyager 2	1979—flyby	1981—flyby	1986—flyby	1989—flyby	
		Galileo	1995—2003 orbiter				
		Cassini—Huygens	2000—gravity assist	2004—2017 orbiter			
		New horizons	2007—gravity assist				2015—flyby
		Juno	2016—orbiter				
	In development	Europa Clipper	2030—orbiter				
		Dragonfly		2034—rotorcraft			
ESA	Launched	Juice	2031—orbiter				

agency to have sent missions beyond the Asteroid Belt and to the outer planets, although the European Space Agency will join this exclusive club when its Juice mission arrives at Jupiter in 2031.

Yet, more wonders were to come. Uranus' moons intrigued scientists with Titania, the biggest, showing signs of cracked ice on its surface and Ariel hosting smooth crater-less plains (see Chap. 9). Another surprise was Triton, the largest of Neptune's moons. Not only did the images returned by Voyager 2 show the moon to have a relatively young surface, but it was also featuring signs of ongoing geological activity as geyser-like vents spewing gases and dark particles were discovered in the southern hemisphere. And all this even though Triton experiences the smallest eccentricity of any known object in the solar system (0.000016), making its orbit almost a perfect circle. Triton is a unique among the big solar system moons, as its orbit is retrograde, meaning that it moves in the direction opposite of the rotation of its parent planet. Such a configuration can only mean one thing: Triton was formed separately from Neptune before being captured by it. Such a capture would have placed Triton on a highly eccentric orbit, generating intense stresses within its interior that might have melted some of the subsurface ice. Could Triton still host underground bodies of liquid water today? Calls to go back to Triton started as soon as Voyager 2 left Neptune and started its journey towards the edge of our solar system.

In the following decades, the success of dedicated orbiters such as Galileo around Jupiter (1995-2003), Cassini around Saturn (2004-2017), Dawn around Ceres (2015-2018) as well as new space probes venturing further out such as New Horizons visiting Pluto (2015), meant that scientific data came pouring in on a yearly basis, providing further evidence to substantiate the claim that liquid water was present within planetary objects; in particular for Callisto, Ganymede, Europa, Triton and Enceladus. And as planetary scientists got ever more enthusiastic about such discoveries, they perfected their theoretical models and other moons in the outer solar system such as Ariel, Rhea, Charon, Oberon, and Titania were found to be potential candidates as well. In fact, it seemed that the outer solar system was awash with liquid water.

But where did all this water come from?

Chapter 2
The Frost Line

The Origins of Water

Like many space-related misconceptions that refuses to go away, there is still a widely held belief amongst the public that our planet is the only place in our solar system where water exists. This couldn't be further from the truth. We find water everywhere.

Break H_2O down into its two main constituents and you get hydrogen and oxygen, respectively the first and third most common elements in space.

Hydrogen, the first and simplest atom in our universe, was formed only 400,000 years after the Big Bang and makes up 75% of all observable matter in the universe. You could refer to it as the primary building block of the universe. In fact, it plays a crucial role in the formation of stars. Hydrogen's abundance is why common elements are found in their hydrogenated forms: oxygen as water (H_2O), carbon as methane (CH_4), nitrogen as ammonia (NH_3), and silica as silane (SiH_4), for example.

Oxygen, on the other hand, was not formed by the Big Bang but was instead cooked inside massive stars. When such stars are born, they initially start fusing hydrogen into helium. However, as the hydrogen in their cores gets depleted and temperatures increase, stars expand into red giants, creating super-dense, super-hot cores. That's when a helium fusion process referred to as the CNO cycle (carbon-nitrogen-oxygen) takes place, forming new elements such as beryllium, lithium, carbon, nitrogen, and oxygen. Since oxygen is a light element with an atomic number of eight, it will be manufactured in great abundance. Later, when the stars are at the end of their lifecycles and have depleted their fuel, these elements and heavier ones get dispersed in vast, interstellar molecular clouds called nebulas.

Because of this, oxygen exists in great quantities. To put this into context, by mass, oxygen makes up 0.9% of the Sun's mass (that is still thousands of times

© The Author(s), under exclusive license to Springer Nature Switzerland AG 2024   19
B. Henin, *Exploring the Ocean Worlds of Our Solar System*, Astronomers' Universe, https://doi.org/10.1007/978-3-031-62953-2_2

greater than the mass of Earth), 49.2% of Earth's crust, and 89% of the world's oceans. Raise your hand and look at it. Two-thirds of it is made up of oxygen.

Back to the nebulas, when atoms of hydrogen and oxygen meet, on the surface of tiny silica grains for example, a simple collision between these two elements provides enough energy for them to combine and create H_2O molecules. Ice is formed. As this happens on a grand scale, significant amount of water lay out there in space.

The Frost Line

In Earth science, where scientists gaze more at their feet than above their heads, there exists a concept called the frost line, which is the maximum depth of ground below which soil will not freeze in winter. This is because our planet's rocky crust is continuously hot regardless of the low temperatures observed on the surface. Below this line, water remains in its liquid form, preserving the organisms located there. Astronomers, being an efficient bunch, poached this term to explain a similar process occurring across our solar system.

But before we go further, let's make ourselves a solar system. Take a giant molecular cloud (nebula) packed with dust (rocks or metals), ices, and sometimes gas, and with enough time, stuff will start to clump together due to the gravitational pull. As overdensities become more massive, large areas of the cloud quickly collapses, and a protostar is born along with various protoplanets.

As these events take place, water particles in proximity to the bourgeoning young star (protostar) will change into to their gaseous form due to higher temperatures they are being exposed to. Further out though, where it is colder, water will continue to exist in its solid form as ice. Thus, in astronomy, the frost line (also referred to as snow line or ice line) is the distance from the protostar where the temperatures are not high enough for volatile molecule (such as water, ammonia, or methane) to change from their original solid state (ice particles) into a gaseous state. For the water molecule, the frost line is a little less than 5 astronomical units (AU), or around 700 million kilometers from our sun, at which point the average temperature falls below 170 K (-103 °C). At this location, between Mars and Jupiter, water in its gaseous state will condense back into ice. (Note that water in its liquid state cannot exist in space due to the lack of pressure.)

One might think that, regardless of the state the water particles are in (gas or ice), they will be used as building blocks for planetary bodies. Thus, the composition of the planets and moons that reside within the frost line, such as Earth, will naturally include such water. That must be the reason why we have water on the surface of our planet. Et voila! Case closed. Well, no, not really. Like many things involving space, the case for water on Earth and other solar system objects is not that straightforward.

This state of volatile particles as either ice or gas is significant as it determines how these particles behave within the solar system. Before the frost line, water is in its gaseous state as vapor, and since it weighs very little, it gets blown away by the intense solar radiation. Actually, the powerful radiation continuously emitted by our Sun will push away any light molecules, small particles and volatile compounds such as water, methane, nitrogen, ammonia, and carbon monoxide. Heavier compounds and elements such as metals or silica (rock) are too heavy to be nudged by solar radiation and remain where they are.

Thus, inside the frost line, there's not much volatile to be seen. However, straight after the water frost line where lower temperatures force water vapor to condense back into ice, newly formed grains of ice pile up and start attracting each other, assembling into bigger chunks. As these chunks get heavier, they become less influenced by the solar radiation and linger there, ready to be used as building blocks by whatever planetary body is being formed in this part of the solar system. A fine example of this is the asteroid belt where the frost line used to lie billions of years ago at around 2.7 AU. The asteroids within the early water frost line have little water content in them (for example Vesta) while those that lie outside of the early water frost line are water heavy (for example Ceres) (Fig. 2.1).

The existence of the frost line explains why we observe a rocky inner solar system and an icy outer solar system as each area of the nascent solar system will contain different condensates for planet formation. You can think of the inner protoplanetary disk, which is the disc of dense gas and dust surrounding a young newly formed star, as a place rich in heavy solid elements, while the outer disk will consist of lighter elements such as ices and gases.

You can easily replicate the concept of the frost line at home by creating the early solar system on your dinner table. Take some salt and pepper and sprinkle them across the table; these will represent the light, volatile compounds such as water,

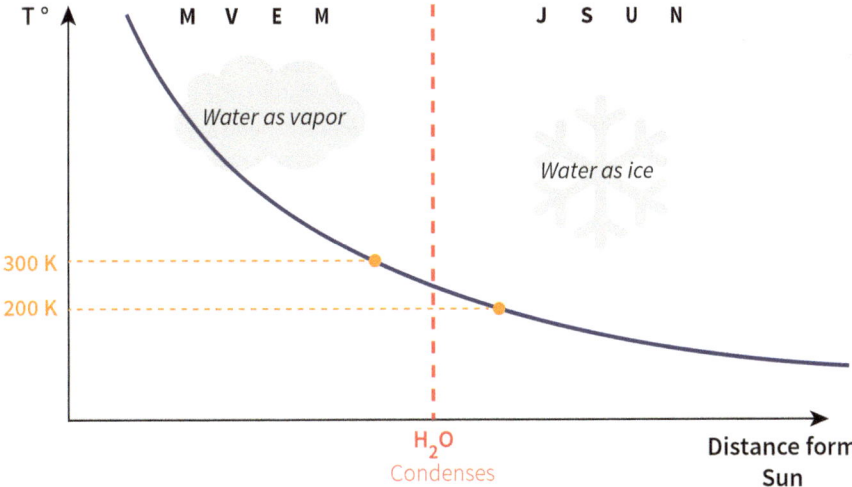

Fig. 2.1 As seen in this simplified diagram, water's current frost line is right before Jupiter at a little less than 700 million kilometers, at which point the average temperature falls below 170 K (−103 °C). At this location, water in its gaseous state will condense straight into ice. The water frost line used to lie within the asteroid belt at the beginning of the solar system.

methane, and nitrogen surrounding a nascent star. Next, go around your home and grab some larger, heavier objects, such as sugar cubes, paper clips, or small pebbles from a garden, and place them randomly on the table. These will represent the heavier compounds present in the protoplanetary disk such as silica and iron as they aggregate into large clumps.

Now, sitting at one end of the table, pretend that you are a newly formed star emitting solar radiation. The way to do this is by blowing as hard as you can on the surface of the dinner table or to use a fan. As you might expect, the heavy compounds will remain on the table. But observe the lighter ones. Where did they go? If the early star within you hasn't blown them all off the table, you will notice a boundary where the lighter elements remain. You've got yourself a salt/pepper 'frost line'.

Now pile the heavy items on the table in little mounds, making sure that you clump them according to the distance of where the wind was blowing, and do the same with the lighter ones. Congratulations, you have just made yourself a proto-solar system!

You will notice that the further you go from where you were blowing, the more you will find lighter stuff in the mounds. You now have displayed variances in the densities we observe in planetary objects across our solar system.

Mercury, Venus, Earth, and Mars, the four inner planets of our solar system, also known as rocky or terrestrial planets, are the 'inner dinner table' planets. They are dense, mainly composed of heavy matter such as rocks and metals and hold relatively few volatile compounds. In contrast, the four outer planets—Jupiter, Saturn, Uranus, and Neptune—are the 'lighter' planets. They are mainly composed of volatile compounds and contain relatively small amounts of heavy matter in relation to their overall masses.

The frost line explains why the 'lighter' gas planets can't form near a star. It also explains why water constitutes a large part of most of the objects lying further from the line, with moons such as Mimas almost entirely made of water while our Moon has essentially none. Giant planets such as Neptune and Uranus hold untold volumes of water within their deep interiors, while Venus is bone dry. The Kuiper Belt, a vast ring of icy objects beyond Neptune's orbit, is populated by billions of icy objects, whereas the Asteroid Belt has some.

You might ask how do we know if objects that are 'light' or 'heavy'? We work this out by measuring an object's density. Density is the mass of the object divided by its volume. We can evaluate the mass of an object by studying how it orbits its parent object, be it the Sun, a planet or smaller bodies such as asteroids, while the volume is directly related to the diameter of the sphere (for spheroid objects, at least) which can either be measured directly if the disc is apparent or estimated through the analysis of the light reflected by the object.

For more than a century now astronomers have been hard at work measuring the densities of the objects in our solar system, to better understand what these are really made of. Frozen water has a density of about 1 g per cubic cm (g/cc or gm/cm^3), while rock is around 3 g/cm^3. Earth's density is 5.5 g/cm^3, which implies that something heavier than rock lies within it: an iron core about 20% of the planet's radius. Surprisingly, Saturn's density is less than water at 0.7 g/cm^3—it is

Table 2.1 Mean densities of the eight planets in our solar system. There is a clear divide between the inner and outer planets. Note that Saturn is lighter than water.

Planet	Mean density (g/cm3)
Mercury	5.4
Venus	5.2
Earth	5.5
Mars	3.9
Jupiter	1.3
Saturn	0.7
Uranus	1.3
Neptune	1.6

so light that it would float in a swimming pool—implying that the giant planet is made up from large amounts of gas. Pluto's density is around 2 g/cm^3, halfway between rock and ice. Its composition, therefore, will be a mix of these two components. As you can see from Table 2.1, when comparing the densities of the four inner planets and the four outer planets, a clear divide between both categories can be seen.

Knowing an object's density therefor offers a powerful insight into its makeup. If there are large pockets of volatiles (e.g., water, ammonia) trapped within a moon or a dwarf planet, we will know. Coming back to the frost line, you will have noticed that, according to this model, objects that have been formed within the water frost line should have none. Yet, Earth is a watery place with 71% of the surface covered. Water also exists in the air as water vapor, in rivers and lakes, in icecaps and glaciers, in the ground as soil moisture and in aquifers. In fact, our bodies are made of 65% water. We also know that Mars, located before the frost line as well, had a very wet past with a vast ocean most likely covering most of its northern hemisphere. Furthermore, it is possible that Venus might also have supported oceans in its past before turning into the furnace that we know. Even more striking is the fact that huge quantities of water-ice have been discovered trapped in frozen pits at the north pole of Mercury, the planet closest to the Sun.

So, why is water found on the inner planets and where does it come from? The simple answer is that we don't know for sure yet, but we have a good idea as to how this might have happened.

The Grand Tack Model

A model referred to as the Grand Tack hypothesis is still a model and not a theory, but it goes a long way in explaining many of the quirks found in our solar system. Here are a few.

Mars as a rocky inner planet should be much bigger than it is now. Why? Because Earth and Venus are the same size, and Mercury was most likely much bigger than it is today because it mainly consists of an iron core with a radius of about 2074 km representing 85% of the planet's radius. The current reason Mercury has such an enormous iron core is that it was once believed to be a much bigger planet before a large object, about the size of Mars, collided with it early in its history. The impact of this collision would have caused Mercury to lose a significant amount of its mantle and crust, leaving behind the large iron core that we see today. Thus, Mars is, based on formation mass, considerably smaller than the other three rocky worlds (Fig. 2.2).

Another quirk is the fact that Uranus and Neptune are very large planets despite being located so far out in our solar system where building blocks were fewer and far between.

The asteroid belt is also an interesting case because it is a mix of rocky and icy bodies. Some of the asteroids appear to have formed further out than the asteroid belt. Regardless of where the frost line might have been at an earlier time, the unusual mix in the asteroid belt suggests there was considerable mixing and stirring up at some point in that region.

Another intriguing aspect of our solar system is the lack of super-Earths: an exoplanet with a mass higher than Earth but smaller than the ice giants Neptune and Uranus. Wherever we look, super-Earths seem to be present in many exoplanetary systems, so much so that some scientists believe that having one is probably the norm and not the exception. If that is the case, where did our super-Earths go?

To answer all these, planetary scientists have been hard at work in the past few decades crunching numbers and working out simulations. One hypothesis that stands out from the others is called the Grand Tack model; another term pinched by

Fig. 2.2 This artist's concept shows the approximate relative sizes of the terrestrial planets of the inner solar system. From left to right: Mercury, Venus, Earth, and Mars. Correct distances are not shown. (Image courtesy of NASA/Lunar and Planetary Institute)

astronomers from an unrelated field. This time, tack is a sailing manoeuvre for changing direction.

The Grand Tack model postulates that Jupiter and Saturn moved inwards and then outwards during the early phases of our solar system, creating havoc in the process. Such a migration is predicted to have disturbed many objects lying within the inner solar system and the outer solar system alike, somewhat mixing them up in the process if not throwing them out of the system entirely which might explain the lack of super-Earths. Furthermore, the Grand Tack model goes a long way in resolving the Mars problem as it shows that Jupiter's gravitational interactions with smaller objects depleted the regions where Mars and the asteroid belt are situated (which might also explain the Late Heavy Bombardment phase). The Grand Tack model also posits that Uranus and Neptune followed Jupiter and Saturn inwards before migrating outwards, scattering in the process icy bodies that had formed in collision-rich regions.

Thus, in the early stages of our solar system, the four water-free protoplanets residing inside the frost line (a bigger Mercury, Venus, Earth, and Mars) were pummelled by small water-rich objects that formed beyond the water frost line.

Until recently, one of the uncertainties with this model was the delivery mechanism for the water. Did water arrive on rocky planets by icy asteroids formed just outside of the frost line or water-rich comets lying further out beyond Neptune's orbit (a long-time favorite candidate)?

An important piece of the puzzle was found on December 10, 2014, when the European Space Agency (ESA) announced the findings of its spacecraft, Rosetta, which had been orbiting the duck-like comet called 67P/Churyumov-Gerasimenko. It was a big day for many planetary scientists, as the data from one of Rosetta's scientific instruments, ROSINA (Rosetta Orbiter Spectrometer for Ion and Neutral Analysis), was made public. Did comets like Rosetta bring water to Earth?

Built by the University of Bern in Switzerland, ROSINA is a mass spectrometer that among other things, analyzes the ratio of heavy water. This ratio, acting as a distinctive signature, allows scientists to determine if Earth's water is similar to water found on extraterrestrial objects such as 67P/Churyumov-Gerasimenko. Before we go into the results, let us clarify the concept of heavy water, isotopes, and the context in which the measurements of 67P/Churyumov-Gerasimenko were made.

The hydrogen atom is composed of a proton and an electron orbiting around it. This form of hydrogen can be referred to as a protium, although this term is rarely used. In some instances, a neutron can join the party and sit comfortably within the nucleus alongside the proton without much fuss. While neutrons do not have electrical properties, they do have mass, and their addition will make the hydrogen atom heavier, thus changing its properties. This variance of hydrogen called deuterium.

Protium and deuterium are called isotopes, meaning that they share the same number of protons and electrons but have a different number of neutrons in their nuclei. In this case, they are the isotopes of the hydrogen atom. These isotopes can be denoted like such: protium as 1H (no neutron), also called Hydrogen-1 and deuterium as 2H (1 neutron), also called Hydrogen-2. As might be expected, more neutrons can be added to the hydrogen atom, thus forming new isotopes: tritium (3H or

Hydrogen-3) containing two neutrons, hydrogen-4 for three neutrons, and so forth, all the way up to an isotope named hydrogen-7 (7H), which unsurprisingly consists of six neutrons. Protium, deuterium, and tritium occur naturally with protium being by far the most common hydrogen isotope in space at 99.98% by mass, followed by deuterium at 0.02%. Tritium is rare and has a very short half-life (12.3 years) while isotopes 4H to 7H are not found naturally as they are highly unstable but can be created in high-end laboratories.

Since the molecule of water, H_2O, contains a hydrogen atom, it follows that there could be multiple 'flavors' of water molecules, each made up of the different isotopes of hydrogen (note that oxygen isotopes also exist, but they aren't relevant here). A water molecule such as $1H_2O$ is made up of protium and is referred to as 'light water,' while a water molecule made up of deuterium, $2H_2O$, is called 'heavy water' and has different physical and chemical properties. Tritiated water, $3H_2O$, composed of tritium is radioactive and only stable for a limited number of years.

The water that we drink every day is a mixture of light and heavy water, although the concentration of heavy water on Earth is extremely low, at around 156 molecules of heavy water per million molecules of water, or 0.0156%. Scientists use this figure as the deuterium/hydrogen ratio, also known as the D/H ratio. It is our planet's water fingerprint. And like fingerprints, each planetary body has a unique D/H ratio.

The mix of these 'flavors' is dependent on the physical properties where the water molecules formed. It is for this reason that astronomers have been busy trying to measure D/H ratios either directly (in situ) or through astronomical observations on a large variety of solar system objects, including planets, moons, many asteroids, and comets. It is important to note, though, that with time, atmospheric or geological processes can alter these ratios. This is considered when analyzing the raw data.

Earth and carbonaceous chondrites meteorites hailing from the Asteroid Belt have very similar D/H ratios in their water and therefore must share the same origin. On the other hand, the four gas planets, as well as long-period comets originating from the Oort Cloud have very different D/H ratios implying that even though comets are thought to represent a significant reservoir of icy material, they are not the bearers of water for our planet.

At first, asteroids seemed to be the most likely culprits, yet, as often in space, things aren't that simple. Short-period Jupiter-family comets such as 103P/Hartley 2 or 45P/H-M-P have also shown to have a D/H ratio very much like Earth's, suggesting that they could also be the contributors to Earth's oceans, contrary to their distant cousins the long-period comets.

It is at this point where the short-period Jupiter-family comet 67P/Churyumov-Gerasimenko comes into the story. Like most comets orbiting close to Jupiter, 67P/Churyumov-Gerasimenko is believed to have originated in the Kuiper Belt. The comet was then nudged inwards, most likely due to Neptune's interaction and caught by Jupiter's massive gravity, forcing it into a short-period orbit around the giant planet (less than 20 years per orbit). A similar pattern must have occurred for short-term comets 103P/Hartley 2 and 45P/H-M-P. Because of this, most scientists expected that Comet 67P/Churyumov-Gerasimenko would display the same D/H ratio as comets 103P/Hartley 2 and 45P/H-M-P and therefore confirm once and for all that short-term comets as well as asteroids were the carriers of water to Earth. ESA even produced in 2014 a short promotional film for the Rosetta mission with this theme in mind (see 'Ambition the film').

But the measurements returned by ROSINA took planetary scientists by surprise. The water ratio of 67P/Churyumov-Gerasimenko was roughly three times higher than that of water on Earth as well as also being very different than that of Oort Cloud comets. According to these results, it seems that the water within Jupiter family objects has diverse origins. For now, because few comets contain Earth ocean-like water, we must postulate that most of the water on our planet was delivered by carbonaceous chondrite asteroids despite their relatively low water content.

Another exciting mission that could shed light on the origins of water on Earth took place recently: the Japanese sample-return mission Hayabusa 2, a successor to the Hayabusa mission (peregrine falcon in Japanese). Launched by the Japan Aerospace Exploration Agency (JAXA) in December 2014, Hayabusa 2's goal was to study the asteroid 162,173 Ryugu, a carbonaceous asteroid that is about 900 meters in diameter.

Hayabusa 2 arrived at Ryugu in June 2018 and spent more than a year studying the asteroid. The spacecraft performed two touchdown operations to collect samples from the surface of Ryugu, the first mission to successfully collect samples from a carbonaceous asteroid. Excitingly, the spacecraft sampled two different parts of Ryugu: one from the surface itself and one below the surface as it had managed to excavate a crater of about 10 m in diameter using a copper impactor, exposing pristine material. The total amount of sample collected by Hayabusa 2 was 5.4 g, equivalent to about a teaspoonful. While that might not sound like much, it is far more than what was estimated during mission design, allowing for thousands of dust grains to be analysed back on Earth. Hayabusa 2 also deployed a rover and a lander to explore the asteroid.

The spacecraft successfully returned to Earth in December 2020, carrying with it samples from Ryugu and in September 2022, the results of 2 years of study were finally published making for a fascinating read. So, what does Ryugu tell us? For a start, the analyses of Ryugu samples revealed strong chemical and mineralogical similarities to Ivuna-type carbonaceous chondrite meteorites; carbonaceous because it has a relatively high percentage of carbon and organic compounds and chondrite because it hasn't been significantly modified by high temperatures. In other words, Ryugu hasn't changed much since it was formed over 4.5 billion years ago.

We also know that it originated from the debris left over after a collision between two larger asteroids, at a time when planets were still forming, although Ryugu's parent body originated from much farther out in the protoplanetary disk. In addition, compounds detected in the samples such as phyllosilicate minerals (phyllosilicates, contain among others, hold water), carbonates, magnetite, and sulfides indicate that Ryugu's parent body has had extensive water alteration in the first few million years of its formation and that liquid water was present. Could this water be the source of Earth's oceans? It doesn't seem so. The water molecules found trapped in the grains are deuterium-rich compared to Earth water, leading scientists to question once more the role of asteroids in delivering water and other volatiles onto our planet.

Another asteroid-sample return mission worth noting is NASA's Osiris-Rex which collected between 400 g and over 1 kg of sample material form asteroid Bennu in October 2020 and delivered them safely back on Earth in September 2023. Bennu is a carbonaceous asteroid as well and there is much hope that water trapped within the sample will provide further insight on the role of asteroids in shaping Earth's volatiles. At the time of writing, results from the sample analysis had yet to be published.

There is no doubt though that the formation of our solar system was complex, and the full picture of the origins Earth water is likely to be more complicated than our current understanding has it and the Grand Tack model, as appealing as it is, has yet to be verified. Another popular model, termed the Nice Model, also provides much insight as to how small solar system objects that formed in the outer solar system were tossed about as Jupiter and Saturn entered a 2:1 resonance, wreaking havoc with the original Kuiper Belt objects. Like the Grand Tack model, the Nice model seems to fit many observations made but remains to be proven. As further D/H measurements are performed in the coming decades through the continued robotic exploration of asteroids and comets (NASA is sending new missions to the asteroids and Jupiter's Trojans), we will hopefully get a bit more clarity on the origins and distribution of water in our solar system.

Note that since water is one of various substances to have a frost line, it is necessary to specify which frost line is being discussed: the water-frost line, the methane-frost line, the nitrogen-frost line, etc. Due to the inherent nature of the water molecule, it is the first volatile substance to condense as we move further out from the Sun. It does so at around 150 to 170 Kelvin (-123 to -103 °C) at roughly 5 AU, shortly before Jupiter's orbit.

As we move further outwards, more volatile compounds turn into icy particles. Nitrogen's frost line is at 63 K (-210 °C), which is roughly within Saturn's orbit, while methane's frost line is at 41 K (-231 °C), before Uranus's orbit. As such, methane ice is present in large quantities on the surface of Neptune's moon Triton and dwarf planet Pluto, giving them a pinkish coloring, whereas surfaces dominated by water-ice such as the icy Jovian moons which are somewhat white or greyish depending on what else has accumulated on the surface. Note that the detection of marginal quantities of ammonia and methane on the surface of Ganymede or Callisto is not implausible as these gases can be trapped in complex molecular structures.

Understanding where each frost-line occurs allows us to extrapolate the composition of the objects formed throughout the solar system. For example, Europa lying within Jupiter's orbit is mainly composed of rocks and water-ice as we are still too close from the Sun for other volatiles such as ammonia and methane to condense. Further out, Titan, the most prominent moon of Saturn, is composed of rocks as well as water, nitrogen ice, and a bit of methane (see Chap. 7). On Pluto, which is located beyond the frost-lines, the surface is a combination of water, nitrogen, methane, and carbon monoxide ices (see Chap. 10) (Table 2.2).

Table 2.2 Abundance of the major ices resulting from the gas nebula. After water, methane and ammonia make up most of the ices.

Ice species	Formula	n_x/n_{H2O}
Water	H_2O	1
Methane	CH_4	0.38
Ammonia	NH_3	0.14
Carbon monoxide	CO	0.054
Hydrogen cyanide	HCN	0.014
Nitric oxide	NO	0.0014
Nitrogen gas	N_2	0.00035

Now that we have a better insight in the origins and distribution of volatiles in our solar system, the topic of our next chapter will look at the life that potentially thrives in them. We will do this by reviewing the most recent notions of astrobiology, such insights will be helpful when we scrutinize in detail each ocean world candidate and make educated guesses as to their ability to host lifeforms.

Chapter 3
Life on Earth and in Space

Although this book is devoted to subsurface oceans of our solar system and the potential for life to arise in such environments, a detour to the barren plains of Mars is recommended as unique experiments related to extraterrestrial life were performed there. As such, the search for life on the red planet is a fascinating tale, and we do it no justice by covering it here in a few paragraphs; entire books have been written on this subject alone. Regardless, such a story provides a sense of perspective before we dive into the depths of the Antarctic ice and with it, into the latest research in astrobiology and the possibility of alien life in the ocean worlds.

Where Are the Martians?

The year was 1976; a year that could have changed our perception of life and ultimately, and maybe our history. It turned out not to be. The robotic emissaries that were sent out to Mars to complete this accomplishment returned ambiguous data, confusing scientists as well as the public that ultimately funded these missions. Disillusionment settled in, and the budget approvers took note. Over the next 40 years, no further mission with a similar scope was flown again and thus ended humanity's first real endeavor into the search for extraterrestrial life in our solar system.

We are, of course, referring to NASA's Viking mission which successfully placed two orbiters and two life-seeking lander at Mars, a remarkable feat in its own right. Much has been said and written about these ambitious missions that seemed, in many ways, far ahead of their time.

The Viking missions were a product of their era, the Cold War, where a powerful combination of national prestige and the need for displaying one's technological prowess was synonymous to bold ideas taking place in space and, to attain them, seemingly limitless budgets. The following figures will speak for themselves.

© The Author(s), under exclusive license to Springer Nature Switzerland AG 2024   31
B. Henin, *Exploring the Ocean Worlds of Our Solar System*, Astronomers' Universe,
https://doi.org/10.1007/978-3-031-62953-2_3

Roughly 1 billion dollars were spent in the 1970s on the Viking program alone, equivalent to a dizzying 11 billion dollars nowadays. To put things in perspective, in fiscal year 2023, NASA received a total of 25 billion dollars (the equivalent of 0.3% of the entire U.S. federal budget) allowing it to fund four different directorates: Human exploration and operations directorate, Aeronautics directorate, Science directorate, and Space technology directorate.

It is within the Science directorate that sits the Planetary Science division responsible for sending robotic spacecraft whizzing throughout our solar system. In 2023, the Science directorate received 7.8 billion dollars of which 3.2 billion dollars were given to the Planetary Science division (other divisions within the Science directorate include Earth science, heliophysics, and biological science). Viewed in such light, the Viking mission with its excessive price tag is an anomaly. The Viking mission can also be considered as a flagship missions. As one can see, flagship missions take a considerable chunk of NASA's annual budget and are therefore few and far between. One such mission was the Cassini-Huygens mission to Saturn which launched in 1997 and ended in 2017, costing in its entirety 3.3 billion dollars. Another flagship mission is Curiosity, the Martian rover, which landed on Mars in 2012 to investigate the planet's past habitability and was done with an overall budget of 2.4 billion dollars, while its newer sibling, the Perseverance rover, hit the 2.7 billion dollars when it arrived on the Red Planet in 2021. Moreover, the coming Europa Clipper mission aimed at investigating Jupiter's icy moon and planned for a 2024 launch had an initial estimated cost of 2.1 billion dollars, although this has now ballooned to 5 billion dollars today much to the dissatisfaction of the U.S. Congress. To put it plainly, NASA would not be provided with the funds to carry out the Viking missions today regardless of their potential to 'change the world.'

Various factors explain why the cost of the Viking mission was so exorbitant. One of them was the doubling down strategy; due to the high failure rate experienced in the early years of space exploration by launch systems and spacecraft, most space agencies would build two spaceships per mission instead of one to guarantee a success. The spacecraft sent to do the first survey of Venus, Mariner 1 and 2, were launched at the same time, while Mariner 3 and 4, 6 and 7, and 8 and 9 were all Mars flyby missions sent in tandem. There were also two Voyager spacecraft and the two Mars rovers Spirit and Opportunity although these were launched decades later. Viking 1 and Viking 2 were not the exception but the norm. Nowadays this mode of operating isn't favored by NASA, in part due to the increasing complexity of space missions, and thus the increase in the cost of a single spacecraft, but also in part due to the higher levels of engineering competency gained throughout the decades. As such, the Jet Propulsion Laboratory (JPL) development center in Pasadena and Goddard Space Flight Center in Maryland, have now produced an uninterrupted succession of overall successful missions over the last two decades, reaching a level of reliability that doesn't justify the need to duplicate spacecraft anymore.

Another factor of the Viking price tag was the space race. In the 1960s and 1970s, the Soviet Union and the Americans were vying to outpace each other technologically; one of the areas pursued was the newly gained ability to robotically explore of the solar system. As such, multiple spacecraft were sent to the Moon, Venus, and Mars. Over the course of a decade, the Soviets launched well over a dozen attempts at the red planet, including flybys, orbiters, and landers. Sadly, apart from Mars 2 and Mars 3 which successfully placed themselves in orbit around Mars in 1971, all other missions failed. As can be seen from Table 3.1, other leading space agencies have sent missions to the red planet ever since, with varying success.

Luckily for planetary scientists, the Americans had more luck. The Viking missions demonstrated the country's technological superiority by being the first to successfully land a spacecraft on Mars—the Russian's Mars 3 did land on the surface in 1971 but failed to operate—the first to send back surface images, the first to study the Martian soil and weather, and of course, the first to investigate the possibilities of life on the red planet, or any planet for that matter (Table 3.2).

Following the success of the Viking mission, the USSR threw in the towel and abandoned any further missions to Mars in the following two decades (the failed USSR Phobos mission launched in 1988 had for primary goal to study Deimos and Phobos, the moon of Mars, and not the planet itself). With the USSR unable to offer serious competition, the American government soon lost interest and the political will to fund more costly missions to Mars quickly evaporated. No new missions to Mars were approved for the next decade and a half.

Table 3.1 This table shows the successes and failures of the world's leading space agencies to send missions to Mars. NASA leads the pack with 19 successful missions throughout 50 years of exploration.

Mars missions successful	Europe	Europe/Russia	India	USA	USSR	China	UAE	
Mission\Agency	ESA	ESA/Roskosmos	ISRO	NASA	Soviet Union	CNSA	UAESA	Grand total
Flyby				3				3
Lander				4		1		5
Orbiter	1	1	1	7	2	1	1	14
Rover				5		1		6
Grand total	**1**	**1**	**1**	**19**	**2**	**3**	**1**	**28**

Mars missions unsuccessful	China/Russia	Europe	Japan	USA	Russia	USSR	
Mission\Agency	CNSA/Roskosmos	ESA	ISAS	NASA	Roskosmos	Soviet union	Grand total
Flyby				1		5	6
Lander		2		1		5	8
Orbiter	1		1	3	2	7	14
Penetrator				1			1
Rover						1	1
Grand total	**1**	**2**	**1**	**6**	**2**	**18**	**30**

Table 3.2 This table shows NASA missions to Mars success rate throughout the years. NASA has had no mission failures in over two decades.

NASA Mars mission success and failures		
Year	Mission name	Successful
2021	Perseverance	Yes
2018	InSight	Yes
2013	MAVEN	Yes
2011	Curiosity	Yes
2007	Phoenix	Yes
2005	MRO	Yes
2003	Spirit	Yes
	Opportunity	Yes
2001	Mars Odyssey	Yes
1999	Deep Space 2	No
	Mars Polar Lander	No
1998	Mars Climate Orbiter	No
1996	Mars Global Surveyor	Yes
	Mars Pathfinder	Yes
	Sojourner	Yes
1992	Mars Observer	No
1975	Viking 1 lander	Yes
	Viking 1 orbiter	Yes
	Viking 2 lander	Yes
	Viking 2 orbiter	Yes
1971	Mariner 9	Yes
	Mariner 8	No
1969	Mariner 7	Yes
	Mariner 6	Yes
1964	Mariner 4	Yes
	Mariner 3	No

There was also another reason why funding proved elusive after the Vikings: the Martians proved hard to find. Indeed, although the results from the Viking experiments proved challenging to interpret, a consensus quickly formed that the missions had been unsuccessful in their attempts to find life. This setback dealt a significant blow to both scientists and the general public, who held high hopes that the Martian subsurface might harbor microbial life detectable by the landers.

It would have been tempting to think that the Viking story was over, but scientists are a stubborn bunch. There existed enough nuances in the results from the life-finding experiments that some doubts were raised, and, unsurprisingly, given what is at stake, some scientists interpreted measurements from one of the experiments as potential evidence that life had actually been detected. If all this seems confusing, that's because it is. To get clarity over the results of humanity's first true quest to search for life in space, let us review the Viking experiments and what they returned.

The Viking Experiments

When in 1971, Mariner 9, the first spacecraft to orbit Mars, took images of a barren and lifeless world shaped only by geological processes, scientists were compelled to admit that if life were to exist on the red planet, it could only be microbial instead of flora and megafauna. Soon after, JPL's director, Bruce Murray, set out to frame the main goal of the Viking mission: "The primary objective of the mission will be the direct search for microbial life on Mars." This didn't stop the American astronomer Carl Sagan from famously joking that lights should be installed next to the cameras of the Viking landers to attract anything that was out there. In case you wondered, there were no lights.

While Viking's main objective might seem straightforward to most, the biology team for this mission was in fact given an incredibly difficult challenge. For a start, no one knew with certainty what the composition of Mars' surface was. The images taken by Mariner 9 provided many clues, but with no imaging spectrometer on board and no spacecraft landing on the Martian surface, there were no real certainties as to what the ground was made of. Would the landers rest on solid rocks or capsize in thick layers of dust? No one could say for sure. Some scientists had even speculated that the surface consistency might be like that of shaving cream, which might sink the landers altogether. Because of this, the biology team had very little to work on, with no definitive answers to basic questions such as: was there soil and if so, what was its composition? Scientists behind the mission had to begin somewhere, though. With the environment so little understood, they decided to go back to the fundamentals of life on Earth.

As a starting point, they knew that detecting tiny microorganisms would prove impossible given the technological constraints imposed by the rigours of space and the spacecraft. Better to focus on the measurable chemical activity that could betray the microbes' presence instead. Yet, sweeping assumptions had to be made as to what chemical activity would be prevalent on the Martian surface.

Making such assumptions was a daunting task. On Earth, the diversity of chemical processes generated by terrestrial microorganisms is great. There are nitrifying bacteria, sulfur-oxidizing bacteria, iron-oxidizing bacteria, sulfate-reducing bacteria (chemotrophs), phototrophic bacteria (generating energy from light), bacteria that use oxygen (aerobic) while others perish in its presence (anaerobic), etc. Scientists have even recently discovered cells that live off pure electricity, bypassing the need for food or chemical reactions altogether. Where should the search for life on Mars begin?

The Viking mission planners started with the common denominator of all life on Earth: carbon-based organic chemistry. Every biological cell on Earth absorbs and expels carbon-based compounds. The reason is simple. Carbon atoms have properties that make them extremely versatile; no other element comes close. They can attach themselves to an endless number of configurations and form long chains, providing an incredible variety of molecules upon which life utilizes to adapt and survive.

With this in mind, it was decided to focus solely on searching for carbon-based chemistry within the Martian surface. Of course, other types of chemistry might exist on Mars, and in that case, they would be genuinely alien, and the life experiments would miss them altogether, but the line had to be drawn somewhere, and as such, markers based on carbon-based chemistry fit with the central principles of the mission brief.

As the mission planners continued their long and arduous decision-making process for the selection of the life-seeking experiments that would go on the Vikings landers, four basic questions emerged:

1. Does anything in the Martian soil exchange gases with the atmosphere?
2. Does anything in the soil release carbon?
3. Does anything in the soil assimilate carbon?
4. Does the soil contain its own carbon-based compounds?

Each one of these questions on its own would not confirm the presence of life, but taken as a whole, if all of them returned positive, then scientists would be able to conclude that a carbon-based biological process of unknown nature was taking place on Mars. They wouldn't be able to see nor characterize whatever process was lurking within the soil, but only confirm through indirect observation that something resembling Earth-based biology was taking place. Such a 'grand-slam' of positive findings might not bring definitive proof that we are not alone, but it would surely fuel a fleet of robotic life-seeking missions, and perhaps, even provide the much-needed impetus to see human footsteps on Mars.

Equipped with these four questions, the Viking team then set about on selecting four experiments, each answering one of the questions.

The first of these, 'Does anything in the Martian soil exchange gases with the atmosphere?', would be answered by an instrument called the Gas Exchange Experiment (GEX). The idea was simple. Some Martian soil would be scooped up by the lander's robotic arm and placed in a hermetically sealed chamber, where it would then be heated and the air within it regularly monitored. Small amounts of water and 'food' to speed up the metabolic processes would gradually be introduced onto the soil. The assumption here is that if Martian microbes did exist, they might be living in a state of torpor due to the glacial temperatures present on the surface (averaging $-55\,^{\circ}C$) and the lack of abundant nutrients.

The 'food' consisted of a broth of vitamins, sucrose, lactose, amino acids, and other organic compounds on which carbon-based microorganisms should thrive. Once the water and nutrients were added, a gas chromatograph, which is an instrument capable of identifying simple substances such as oxygen, nitrogen, carbon dioxide, and methane, were to analyze the atmosphere in the chamber on a regular basis. Any changes in the atmosphere's composition resulting from metabolic reactions taking place within the soil would be quickly detected.

The decision to introduce food onto the sample of Martian soil to was not universally accepted though, and only two of the experiments would use such a supplement in their processes: GEX and LR (see below). One of the primary concerns was that organic nutrients could create false positives by triggering life-like reactions with existing inorganic compounds. Thus, a positive result from GEX on its own would be difficult to interpret, which is why the fourth experiment, 'Does the soil contain organic compounds?', was crucial to the entire biology package, as it would validate any positive results; in other words, the link between the organic-like reactions detected when the food is introduced onto the Martian soil and the organic compounds detected within the soil is a highly plausible one.

The second question 'Does anything in the soil release carbon?' would be answered by the Labeled Release (LR) experiment. This experiment was designed to detect any signs of organic processes coming from the Martian soil as water and nutrients containing carbon-14 atoms would be added. If Martian microorganisms metabolized the nutrients with the traceable carbon-14, they would theoretically release these carbons as waste gas into the atmosphere of the container. Any gas laced with carbon-14 could then be easily detected by Geiger counters, thus providing indirect evidence that some organic process had taken place within the soil.

To answer the third question, 'Does anything in the soil assimilate carbon?', the Pyrolytic Release (PR) experiment, also known as carbon assimilation experiment, was conceived to detect signs of life in the complete absence of water and organic nutrients as no food would be added. Instead, the assumption taken by Norman Horowitz, the leader of a research project, was that if life was already present on the Martian surface, it should be able to metabolize without the addition of liquid water or food, whose introductions could add unnecessary complexity to an already complex experiment. Furthermore, continuing with this assumption, it could be that the addition of relatively large quantities of liquid water and nutrients onto the soil might potentially stress organisms that have evolved to live in a more extreme environment.

Therefore, for the PR experiment, Martian soil would be collected by the robotic arm and placed into a chamber where only carbon monoxide and carbon dioxide containing carbon-14 atoms would be added to the air inside the chamber. While designing the experiment, Horowitz had found that ultraviolet light could create certain organic compounds in the presence of carbon monoxide, water vapor, and certain types of soil. With this in mind, he filtered out all ultraviolet light created by the arc-lamp meant to simulate the Martian day in the chamber.

The next stage would be to slowly heat up the sample and after 120 h of incubation, the atmosphere within the chamber would be pumped out using helium (a neutral gas) and the remaining soil subjected to temperatures of about 625 °C to pyrolize (break down) any organic matter present into volatile compounds that could be detected. Any traces of carbon-14 in the sample after pyrolization would indicate that something in the soil had captured the carbon monoxide or dioxide in the atmosphere and used it for its processes.

Finally, the last question, 'Does the soil contain its own organic compounds?', would be answered by the Gas Chromatograph/Mass Spectrometer (GC/MS) experiment. Martian soil would then be placed into one of three sample ovens and heated to 500 °C which could be reached in only 8 s, breaking down in stages any organic compounds from a hypothetical Martian microorganism present in the soil.

The resulting vapors generated would be filtered through a gas chromatograph, which acted like a granular filter separating out the organic molecules according to their complexity; the simpler, lighter molecules would pass through the chromatograph at a faster pace than the heavier, more complex ones. The resulting gases would then go through a mass spectrometer which is a sophisticated instrument that bombards the passing gases with charged particles, ionizing the organic molecules in the process. These molecules, in turn, would then be subjected to a magnetic field that deflected them according to their masses: the lighter organic molecules being more influenced than the heavier, more complex ones. Finally, electronic detectors would measure the degree of deflection for each molecule and the results compared with similar experiments done on Earth using organically derived vapors. The GC/MS was a unique state-of-the-art piece of engineering. That such a complex instrument was used on the Martian surface so early within space exploration is remarkable.

In theory, the GC/MS experiment would be the only one to directly detect organic compounds, while the GEX, PR, and LR tests would follow a metabolism-detection approach. It is important to note that the Viking GC/MS wasn't meant to detect life but instead, provide additional data to interpret the other three biological experiments.

And so, it was with these four experiments that the first mission to investigate the possible existence of extraterrestrial life on another solar system body was launched. Of course, the Viking mission was not all about searching for life. It would also take the opportunity to study the surface geological composition, the weather and the atmosphere with additional scientific instruments placed on the landers and the orbiters (Fig. 3.1).

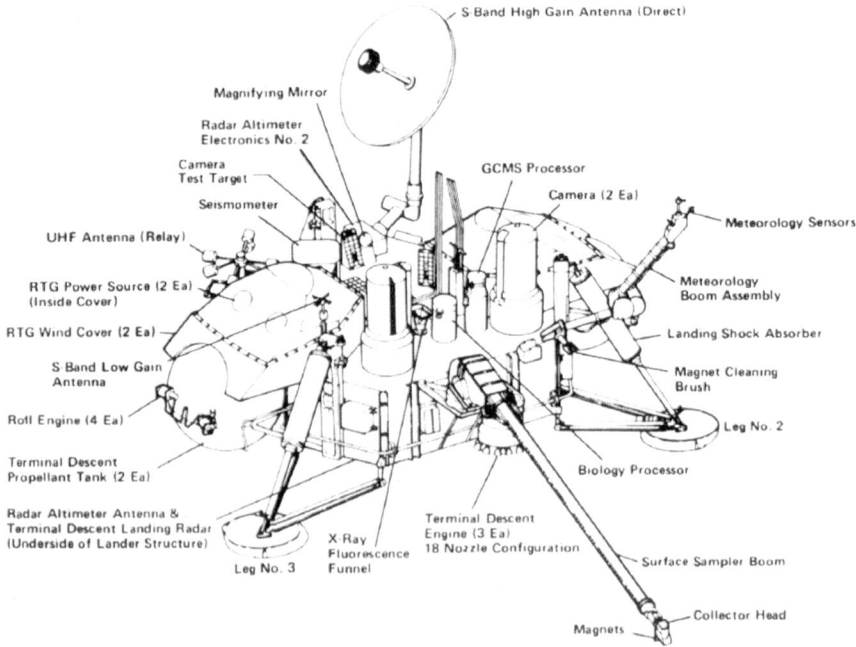

Fig. 3.1 The Viking lander. (Image courtesy of NASA)

Once the mission was approved in 1972, more than 40,000 pieces were assembled per lander—with some pieces miniaturized—and all rigorously tested to withstand the harshness of space travel, making the landers the most complex spacecraft ever built at the time. Hundreds of leading scientists, engineers, and planners worked tirelessly to meet the launch date of 1975. No space agency today would contemplate setting up such an ambitious, costly, and complex mission within the tight schedules imposed on the Viking team.

Geopolitics were to blame for such a situation as the Soviet Union had already sent 13 missions to the red planet by the time the Viking mission was being developed and there was fear that they would send more. The fact that all these missions were unsuccessful mattered less as all that was needed was a stroke of luck and the Russians might be able to hit a first on Mars. Yet, unbeknown to the Americans at the time, the Russians had run out of steam and no further missions to the red planet would be launched for the next 15 years.

Alas, despite the vast resources allocated to the Viking missions, it was inevitable that such a tight schedule would have a detrimental impact on its preparation. For example, there was no time to calibrate all four experiments with each other, meaning that the data collected by the suit of instruments might be difficult to interpret. Also, due to technical constraints, the GC/MS experiment would analyze entirely separate soil samples from the one feeding the GEX, PR, and LR experiments, making the comparison of the results even more prone to ambiguity. And, of course, there was no time to send a preceding mission to analyze the Martian soil and determine in advance its chemical composition and general properties which would have significantly helped the scientists to tailor their life-detecting experiments according to the Martian environment. The clock was ticking, and any mission landing on Mars was still preferable to none in the eyes of many.

All that the mission planners had to do now was to hope that at least one of the landers would come to rest safely onto the surface, and all four experiments would work as designed. At this early stage of space exploration, where more missions to Mars had failed than succeeded, it would be a remarkable feat if such a complex mission would succeed entirely. And indeed, the mission had its share of hair-raising moments.

To find the best landing sites for both landers, the Viking team had scrupulously analyzed all 7000 images of the Martian surface taken by Mariner 9 in 1972 (the first spacecraft to orbit another planet). Yet, when the Viking 1 orbiter inserted itself around Mars, on June 1976, and started to send back high-resolution images of the selected landing sites, the Viking team quickly realized that these sites were not appropriate, and with less than a month to go before landing taking place, hurriedly searched for new locations using the orbiter images.

Thankfully, they found one in time for the Viking 1 lander which safely touched down on July 20, 1976, in western Chryse Planitia hundreds of kilometers north of its intended primary site. The Viking 2 lander touched down on September 3, in Utopia Planitia, a third of the way around the planet from its initial landing site. If more time had been given to review the new images, even better landing sites might have been found for both landers.

Regardless, when Viking 1 safely touched down—making it the first American spacecraft to land on another planet—and Viking 2 did the same shortly after, there was immense relief among the scientific community, the public, as well as the politicians who had funded this mission. America was winning.

The Viking Results

As expected, the first images of the barren Martian plains sent back from the landers showed no lifelike plants amongst the rocks and no hard-shell bug running in the foreground. Mars' surface was bleak and lifeless (Fig. 3.2).

Once the landers were confirmed to be operational, the experiments started straightaway. Lasted weeks at a time, the experiments were nevertheless slowed down due to various technical problems encountered by the landers. For example, on July 22, 1976, the day after the Viking 1 lander touched down, the robotic arm became stuck in the stowed position. Engineers were able to free the arm after several days of troubleshooting, however, it was never able to operate as smoothly as it did before the jam. Also, a radioactive gas leak within the chamber used by the PR experiment was detected on Viking 1, which required adjustments in the data to compensate for the loss. Remarkably though, and despite the problems stated above, all the instruments worked out pretty much as was expected on both landers and the experiments were carried out as planned. From an engineering point of view, the Viking mission with its two orbiters and landers was a resounding success.

Alas, the results from the Viking experiments proved to be confusing for the public and the scientists as only one of the three biological experiments yielded a seemingly positive outcome, while the GC/MS instruments detected no organic compounds. It seemed that the Martian soil wasn't ready to give up its secrets yet. Let us review these results in more detail.

Fig. 3.2 Taken by the Viking 1 lander shortly after it touched down on Mars, this image is the first photograph ever taken from the planet's surface. It shows an arid and rocky world. (Image courtesy of NASA)

On both landing sites, the GEX experiment (Does anything in the Martian soil exchange gases with the atmosphere?) started first and immediately experienced a sudden surge in oxygen and carbon dioxide when the nutrients and water were introduced to the soil—although the experiment on Viking 2 detected only a quarter of the oxygen measured with the Viking 1 experiment. The quick rise in oxygen and carbon dioxide seemed too sharp and sudden for a biological explanation and even more troubling for the 'pro-lifers' camp, both gases tailed off after 50 h of exposure to the fresh food and water.

Such a result was in contradiction with our understanding of life. If Martian microorganisms were present in the soil, why would they suddenly stop feeding on the nutrients we were sending them? Instead, as any chemistry teacher will tell you, the quick bursts and eventual drop in oxygen production had the tell-tale marks of a chemical reaction and a culprit was quickly advanced to explain the GEX result: peroxides. These are molecules, such as hydrogen peroxide (H_2O_2), which are highly reactive with water and metals and release oxygen as a by-product of such reactions. The peroxide hypothesis wasn't universally accepted by the scientific community, but it provided a potential non-biological explanation for what was being observed, and as such, GEX was deemed to have not detected life-signs.

The results from the PR experiment (Does anything in the soil assimilate carbon?) proved unsatisfactory as well. Minute traces of carbon-14 were picked up by the Geiger counters on both landers after the gases were introduced and the soil samples pyrolyzed, implying that something in the soil had captured the carbon-14. So far, so good. Yet, if this was indeed the tell-tale sign of life, quick calculations showed that the level of carbon-14 detected after pyrolyze represented less than a thousand bacteria cells per sample if these cells were using the radioactive carbon dioxide or carbon monoxide as part of their metabolic processes, making it an unusually small number of microorganisms if this turned out to be true. To put this figure into perspective, a single teaspoon of garden soil back on Earth can hold up to a billion bacteria.

This result wasn't direct proof that life was present on Mars, but it was indicating that something was going on. However, bad news for pro-lifers was to come as control soil samples which had been sterilized by heat before they were put in contact with the radioactive gases also proved to contain minute traces of carbon-14, albeit at levels much smaller than the ones detected in the non-sterilized soil samples. If any life was present in the scooped soil, it should have been destroyed by the sterilizing process, and no trace of carbon-14 should have been detected at all. These results implied that a non-biological explanation, such as an unknown chemical reaction, was most likely the culprit. PR was also deemed to have not detected signs of life.

Of all the experiments that were run by the Viking mission, the LR experiment (Does anything in the soil release carbon?) proved to be the most promising in showing lifelike results. Once the water and nutrients were in contact with the soil in the sealed compartment, the Geiger counter in both Viking landers were able to detect carbon-14 in the air. The food had been broken down. The Viking 1 experiment measured high levels of radioactive molecules in the air over the course of

7 days, while Viking 2 measurements went even higher. This startled Gilbert Levin, the project scientist behind the LR experiment.

To remove the possibility of getting positive results associated with the effect of ultraviolet radiation hitting the Martian surface, the landers used their robotic arms to collect soil underneath a rock, which again tested positive. Levin then set up a few more control tests. One test had the soil sterilized at 160 °C, which produced no results. He then heated another sample at 50 °C and found that there was much less activity. Finally, a sample was kept in the dark for 2 months at 10 °C. This returned a negative result. Whatever was going on in the Martian soil had been damaged by extreme heat and lack of sunlight.

Following such results, Levin and his team claimed that the LR experiment had found evidence of life in the Martian soil and that such life had absorbed the carbon-14 before releasing it through a metabolic process. The scientific community wasn't convinced by the results though and suggested non-biological explanations such as oxidants similar to the peroxides that had already been proposed as a possible explanation for the GEX results. As a reminder, oxidants are molecules that can react to organic matter present within the nutrients and break it down to release the volatile compounds containing the traceable carbon-14. However, there were various problems with such suggestions and the case was still open for a biological origin in explaining the LR experiment.

It was the results of the final experiment, the GC/MS (Does the soil contain its own carbon-based compounds?) that turned out to be the most intriguing. As a reminder, if the GC/MS's results failed to detect carbon-based compounds in the soil, then none of the positive results from the three other experiments would hold up.

The Viking scientists were eager to examine the GC/MS data as soon as it was received. As it turned out, the results of the GC/MS experiment became the most contentious set of data of the mission: no traces of organic material were detected within the soil samples taken from both landers. The GC/MS instruments were working—they had detected the cleaning solvents used to disinfect the spacecraft prior to launch—yet, despite numerous attempts, no organic compounds were detected. This was a major surprise to everyone as organic compounds are commonly found on meteorites, asteroids, comets, and numerous other planetary bodies (even in vast interstellar clouds of gas). So, it was expected that the Martian surface would also contain traces of such compounds, albeit at lower levels than what would be found if life was present. The fact that the GC/MS returned blank was a problem.

As the year ended following months of experiments, the mood in the Viking team soured. While the LR results seemed promising, the GEX and PR results didn't support the life-on-Mars hypothesis and questions were being raised regarding the reliability of the GC/MS instruments. This set of ambiguous data resulted in different interpretations of what the Vikings had found on Mars. Levin claimed that his experiment had found life-signs, and that the GC/MS experiment wasn't sensitive enough to detect the organic compounds related to such hypothetical life, while in the other camp, scientists claimed that Levin's experiment didn't discover Martian life but a new form of non-biological reaction. Although, what this was proved difficult to determine in part due to the unfamiliarity with the Martian surface. The teams were divided.

NASA was in a muddle and quickly released an official statement claiming that the results neither proved nor disproved the existence of life on Mars—hardly an acceptable outcome for the lawmakers who had agreed to fund billions of dollars for the mission.

With the public disappointed and scientists split into camps, Washington took note. In the coming years, the budget for the planetary science division evaporated like water in a desert. There would be no further spacecraft sent to the red planet for the next 20 years and no life-detecting mission sent on another world until the Perseverance rover launched in 2020.

In 2007, Klaus Biemann who designed the GC/MS experiment stated: "The thing that gets me annoyed is that people think we were looking for life … Of course we weren't looking for life! We were looking for organic matter." In other words, the GC/MS didn't seem to be sensitive enough to detect the presence of living organisms. Bremen continued by saying that the search for life on Mars is a very difficult and challenging task and that he did not believe that the GC/MS was the right instrument for the job.

And yet, the outcome could have been much different. A less complicated experiment was initially proposed to be part of the Viking biological package but failed to be selected for budgetary reasons. Designed by Wolf Vishniac, an American microbiologist, the proposed experiment, named the Wolf trap, consisted of suspending a sample of Martian soil in an aqueous solution that contained nutrients. The idea was simple: if Martian microorganisms were happy to grow in the solution, they would turn it cloudy as well as change its acidity. Place a light sensor and a pH meter onto the solution, and you've got yourself a life-detector. The experiment wouldn't provide information on the exact nature of the life found, just that a life-like thing has grown in the solution.

Vishniac tested his Wolf trap with many samples from Earth and showed that for some strains of microorganisms, it was a surprisingly effective means of detecting life forms. The only caveat was that by using a fully aqueous setting, this experiment was the least 'Mars-like' environment the Martian soil samples would be subjected to. It didn't seem to matter. Trials done using Antarctic soil which were extremely arid and cold—similar to the conditions expected on the Martian surface—and supposedly sterile of life, returned positive results from microorganisms unknown at the time. The Wolf trap had discovered extremophiles. It seemed that rather than stemming the growth of these lifeforms which had evolved an ability to withstand an extreme environment, the Wolf trap was doing the opposite, and by accelerating their metabolism. After all, the chemistry of life on Earth is an aquatic chemistry; we lug huge amounts of salt water around with us.

It is worth mentioning here that not all extremophiles experience growth within such a water-saturated medium. Therefore, the snag in using such an experiment is that a negative result on Mars would not necessarily imply that alien life within the Martian soil doesn't exist, just that life capable of growing in such medium isn't present within the sample. Nevertheless, because of the simplicity of the Wolf trap design—all you need is a container, some amino acids, and water—and its detectors, astrobiologists such as Charles Cockell, director of the UK center for astrobiology, still see merit in this experiment.

Sadly, the Wolf trap never flew onboard the Vikings or any other spacecraft. It is anyone's guess as to where we might be now if the Wolf trap experiment had been selected for the Viking biological package as was initially planned.

Post Viking Blues

Ever since the LR experiments, researchers have been searching for other kinds of nonbiological chemicals that might produce identical results. In addition, more landers and rovers have added to our scientific understanding on the conditions found on the Martian surface and it has become clear that the Viking missions, ambitious as they were, occurred far too early within the context of the exploration of Mars.

For the nonbiological camp, the discovery in 2008 of perchlorates in the Martian surface by NASA's Phoenix mission which had landed in the northern polar region might begin to explain the LR experiment results: a chemical reaction with the perchlorate and the nutrients would take place within the sample, releasing molecules laced with carbon-14. Furthermore, by-products of perchlorate, chloromethane and dichloromethane, were detected by the Viking instruments suggesting that perchlorates must have been present in the soil samples.

On the other hand, the pro-life camp claimed that those same perchlorates would destroy all organic compounds contained within the soil sample when heated, thus preventing the GC/MS experiment from detecting them. This might also be an explanation as to why GC/MS returned blank. Pro-lifers also point out that none of the strong oxidants proposed over the years by the non-biological camp exhibit the thermal profiles as established by the LR experiments. Given all this, it cannot be ruled out that Gil Levin is correct and that there are dormant life-forms in the Martian soil.

In 2012, Joseph Miller, an associate professor of cell and neurobiology at the Keck School in the United States, and mathematician Giorgio Bianciardi, of Italy's University of Siena, used a data analyzing technique called cluster analysis which groups together similar-looking datasets, and reviewed the data returned by the LR experiment. The computer program analyzed the raw data and established that the results of two of LR's experiments were clustered in the biological group while the rest were in the non-biological group. The same Joseph Miller had previously published a paper putting forward the possible detection of circadian rhythms in the results from the LR experiments. These rhythms are internal clocks present in every life form on Earth that help them time their biological processes such as waking up or sleeping as well as temperature regulation. Miller's interpretation requires further work to be validated.

A breakthrough came in 2014, with the detection by the Curiosity rover of organic molecules in Gale Crater; the first detection of such molecules on the Martian surface. Curiosity was using a new method to analyze the soil which prevented the perchlorates from being involved in the process as opposed to the Viking experiments, and it didn't take long for the rover to detect organic compounds. This discovery raised further doubts on GC/MS's ability to detect organic compounds on the Martian surface as the simple act of heating up the samples prior to the analyses was most likely the cause for the failure to detect any organic material during the entirety of the mission.

Further data analyses and discoveries will undoubtedly continue to keep the results from the Viking missions relevant in the years to come with the Mars Sample Return mission, currently being designed by NASA and ESA, the most likely mission to shed more light on the Viking results.

In the immediate post-Viking era though, the search for extraterrestrial life in our solar system took a step back from such bold beginnings. In fact, the whole planetary science field at NASA went through a 'lost decade' as budgets were slashed in half and no new planetary science mission was launched throughout most of the 1980s.

Finally, in the early 1990s, an increase in funding allowed NASA to contemplate new missions to Mars, although this time, life-searching missions were completely off the table. Instead, a program was devised to make a comprehensive step-by-step study of the Martian geology and its atmosphere; the Mars Exploration Program (MEP) was born. MEP would benefit from a sustained fleet of orbiters, landers, and rovers in the next three decades, with a special emphasis on a strategy called 'Follow the water'.

This approach culminated in 2012 by the landing in Gale Crater of the highly specialized Curiosity rover, which has for its main mission the understanding of the geological and geochemical processes on Mars to determine if it could have supported life. Such a concept is often referred to as habitability, a term widely used in geoscience and defined as the ability of an environment to support the activity of at least one known organism. This concept can be further broken down into either instantaneous habitability, which are the conditions at any given time in each environment required to sustain the activity of at least one known organism, or continuous planetary habitability, the capacity of a planetary body to sustain habitable conditions on some areas of its surface or within its interior over geological timescales.

In other words, Curiosity was to seek out past or potentially present habitable environments in Gale crater with the idea to better characterise the complex Martian environment before implementing missions investigating the plausibility of past or present microbial life.

After years of studying Gale Crater, we now know that at some point in Mars' distant past, it was a habitable place for carbon-based life as we know it. Whether any life forms were present is another matter altogether. This is where Curiosity's sibling, the Perseverance rover, comes in.

Designed and built on the back of the Curiosity rover, the Perseverance mission has for objectives to search for ancient Martian environments capable of supporting life, collecting, and storing soil samples of the most promising surface material, testing an oxygen production device for future crewed missions, and looking for past microbial life in these environments. After decades of probing and sensing the red planet with orbiters, landers, and rovers, space agencies are once again confident in funding life-detection missions to Mars, thus continuing the quest first started by the Vikings 50 years ago.

The life-detection instruments carried by Perseverance bears no resemblance to the ones brought to Mars all these years ago. There are no nutrients to give away and no water to be drizzled onto the Martian soil. Instead, the rover uses, amongst other tools, powerful imaging instruments known as SuperCam on the top of its mast. These include an X-ray fluorescence spectrometer and a Raman spectrometer that can quickly provide chemical and mineralogy analysis of any surface by zapping rocks with the help of a powerful laser. The aim here is not to look for the hypothetical metabolic reaction in the presence of external stimuli but to carry out the hunt for specific molecules present within the soil or rocks and detect chemical patterns in the mineralogy which could betray a biological origin.

To support these spectrometers, Perseverance boasts no less impressive imaging tools positioned at the end of its two-meter robotic arm: PIXL and SHERLOC. PIXL stands for Planetary Instrument for X-ray Lithochemistry and is an X-ray spectrometer coupled with a powerful camera; the Micro Context Camera (MCC). PIXL's X-ray spectrometer provides the elemental composition of a target with a large range of detectable elements (+26 elements) at a high level of sensitivity (at the ten parts per million level) and at fast speeds (5 s per measurement on average). At such high levels of precision, PIXL uses a laser and MCC to calculate, in microns, its distance from the target which a computer then uses to guide a six-legged supporting device, named the hexapod, to aim the X-ray at the target at precisions of just a hundred microns.

Such capabilities allow PIXL to create, from thousands of individual measurements, a three-dimensional map of a surface's composition. It can study fine-scale structures such as grains, veins, laminae, cements, and concretions; in doing so, the instrument looks for changes in texture and chemistry induced by past life such as the presence of mineralized biofilms.

SHERLOC, standing for Scanning Habitable Environments with Raman & Luminescence for Organics & Chemicals, is at its core a UV laser and a Raman spectrometer supported by two cameras: ACI for Advanced Context Imager and WATSON for Wide Angle Topographic Sensor for Operations and eNgineering. As the UV laser hits a target, molecules that contain rings of carbon atoms have a distinctive glow which can reveal the structure of the carbon atoms at microscopic scales and therefore pick up traces left by hypothetical past carbon-based lifeforms.

In addition, Raman scattering provides insight into the mineralogy of the target including the nature of organic compounds. ACI allows closeup views of the target at resolutions of 10.1 microns. WATSON, an upgrade from the MALHI camera used on Curiosity, also takes microscopic close-up images at resolutions of 15.9

microns per pixel. WATSON is there to provide context for the spectroscopy measurements and complement ACI and PIXL's investigations. If the rover stumbles upon signs of fossilized microbial life, it will be ACI, PIXL, and WATSON which will image these at close range; never have a set of powerful imaging systems been sent on another planetary body.

Perseverance's mission is located at Jezero crater. Initially thought to have been a lake at two separate times in the past, recent data returned by the rover seems to indicate that liquid water wasn't particularly abundant in the crater as was initially hoped. Hopefully, all 38 half-inch-wide core samples taken by the rover in the coming years will be collected and sent back to Earth with the Mars Sample Return mission in the late 2030s, although recent problems with the mission design have cast a shadow on its feasibility. Mars Sample Return is a joint mission involving NASA and the European Space Agency (ESA) which has also designed and built its own life-detecting rover named Rosalind Franklin (previously known as the ExoMars rover). Similarly to the Perseverance mission, Rosalind Franklin's mission will be to search for ancient microbial life on Mars and will have, a novelty, a drill to sample the soil within the subsurface. The rover hosts a suit of high-resolution imaging instruments and spectrometers to study surface mineralogy, some of which are the Close-Up Imager (CLUPI) which will be able to take sub-millimetre resolution images of the targets during the drilling operation while an infrared spectrometer (MA_MISS) will determine the composition of freshly exposed material. Sadly, Rosalind Franklin's launch in July 2022 was cancelled for geopolitical reasons and is now planned for 2028.

The search for life on Mars is a fascinating tale—we do it no justice by covering it in just a few pages here—and it provides insights as we speculate throughout this book about the possibility of extraterrestrial lifeforms existing deep within the icy bodies of our Solar System and the means of detecting them. As such, future life-detecting missions sent to the ocean worlds will take a very different approach to the Vikings, defined by the technological and scientific limitations of their time, and will benefit greatly from the expertise gained by missions characterizing the past and present habitability of the plains and craters of Mars. Already, one of Europa Clipper's instruments called Mapping Imaging Spectrometer for Europa (MISE) was designed from the Compact Reconnaissance Imaging Spectrometer for Mars (CRISM) on the Mars Reconnaissance Orbiter. More cross-fertilisations are to be expected.

What Is Life?

Before we go into further detail on the ocean worlds, their potential habitability, and the future missions planned to study them, there is one aspect of the Viking mission that we need to delve into which is a surprisingly tricky exercise: we need to define life. Indeed, the Viking scientists worked hard to come up with a broad definition of life which they framed in terms of something producing a metabolic reaction; the

GEX, LR and PR experiments were set up this way. However, as we shall see, this definition is lacking, so let's go back to the basics.

It is often remarked that biology is the only science in which there is no general agreement on the object of its study. What is life? How should we define it? This simple, yet profound question seems much harder to answer than it sounds. Is a virus alive? A prior? Both can replicate and evolve. For that matter, how about the strawberry you picked up at your local store or from your garden? As it is lying there in your bowl waiting to be eaten, is it alive?

Go to your nearest university and talk to the scientists in the biology department and you will most likely find that there is no simple consensus on these questions. In fact, it turns out there is no official definition of life as it is not a scientific term to begin with. It is instead a popular term, similar to the words planets and continents, which scientists struggle to come up with a satisfactory definition.

So, where do we start? There are various ways to approach this thorny question. Some scientists list characteristics to define life while others prefer focusing on processes.

Throughout the years, though, some attributes related to life have been put forwards:

- Complexity: life seems to bring about organization.
- Metabolism: chemical reactions occurring in an organism.
- Homeostasis: the ability to regulate one's internal environment.
- Growth: the ability to increase in size and complexity.
- Holds information: contains a system for storing data (DNA or RNA).
- Reproduction: the ability to multiply sexually or asexually.
- Responsive: reacts to external stimuli.
- Evolving: capable of Darwinian evolution.

Most of these aren't genuinely satisfactory, though, as various non-biological processes can replicate them while some biological ones can't. For example, a snowflake can be seen as having complexity; crystals can even experience growth if the conditions are right. Fire seems to reproduce as it jumps from tree to tree in a forest fire, while computer programs can evolve on their own, yet no one would consider them alive (at least, not yet). On the other hand, the fact that your neighbor's neutered cat can't reproduce or that a person is not reacting to external stimuli due to a temporary coma doesn't imply that they are not alive.

A definition of life which has gained much attention amongst scientists was proposed in 1992 at an Exobiology Discipline Working Group set up by NASA and comprised of various academics. The definition they came up with is the following: "Life is a self-sustaining chemical system capable of undergoing Darwinian evolution." As we shall see, each term in this definition has been carefully chosen.

The terms "chemical system" supports the notion that the transformation of matter is the bedrock on which life exists. We can't have life from nothing. The expected arrival of artificial intelligence in the coming future, which according to this definition would not be considered as life, will most likely require some changes to this term. Given the context of this book, the definitions discussed here will be from a biological point of view only.

The terms "Darwinian evolution" implies many attributes often associated with life such as self-replication or reproduction, mutability, heritability, and metabolism without which you can't replication. Darwinian evolution also entails adaptability properties such as locomotion, photosynthesis, chemosynthesis, energy storage, etc. Furthermore, Darwinian evolution is a way for complex entities to maintain themselves, not as individuals but as a system continuously evolving to adapt to environmental change.

The term "capable" brings about the idea of the population or living system in contrast to the individual, as many individuals are needed to make the system capable of Darwinian evolution, not just one.

Finally, we have "self-sustaining" implying that all the information necessary for a system to undergo Darwinian evolution must be present within the collective system. In other words, the system should work without the need for an external factor. With such definition, a virus isn't alive, as it doesn't have all the information required for it to undergo Darwinian evolution. A virus within a host cell might be called alive, but this requires an external intervention (the host cell).

This definition is not without its critics. Some scientists argue that it is too broad while others argue that it is too narrow. Despite these criticisms, NASA's attempt at defining life is a useful starting point as it seems to be very close to what we should look for.

To demonstrate how this definition can help scientists searching for evidence of extraterrestrial life, here is an interesting thought experiment proposed by Lin Chao, professor at the University of California at San Diego, using a re-enactment of the Labeled Release (LR) experiment from the Viking mission. As you may recall, the experiment added water and nutrients labeled with C-14 to the Martian soil, which in turn released labeled gas in a lifelike manner. By focusing primarily on metabolic reactions, the LR experiment was inadequate on its own to confirm the presence of life since inorganic processes could have mimicked those reactions and returned positive results.

To overcome this, Chao suggests that the LR experiment be modified to focus on other attributes of life as defined above such as reproduction and evolution. Surprisingly this can be easily implemented.

Let us start with reproduction. The experiment begins as initially designed by placing a sample of Martian soil into a container. Water and labeled food are added, and as we have seen, a reaction occurs, and gas is released. This is where Levin's experiment stops, and Chao's experiment begins: a small portion form the soil sample that has just been tested is now transferred into a new container containing additional food and water. Once again, detectors monitor the air. If the agent causing the reaction is biological in origin and is still active within the soil, labeled gas would be emitted once again, with results equivalent to the original experiment if sufficient time is allowed.

If, instead, the causative agent is inorganic, the amount of gas produced in the second container should be less than in the original experiment. The repeating of this process could even reveal predictability in the amount of gas reduced as the original sample gets diluted.

Chao's assumption here is that such a dilution will not take place as the biological agent will compensate by reproducing and will, with time, be able to generate similar levels of released gas regardless how many times the soil is transferred into a new container.

This new experiment, simple and effective, might still fall prey to false positives as autocatalytic reactions have the ability to mimic growth patterns.

Autocatalytic reactions occur when a compound 'A' comes in contact with a substrate 'B' and produces a reaction that generates twice as many 'A's and a by-product 'C.'

$$A + B => 2A + C$$

Various non-biological autocatalytic reactions are known; fire is the most famous of all. Here, the 'C' by-product could be the release of gas, therefore mimicking the reproduction pattern of Chao's LR experiment even if the experiment starts with a smaller amount of 'A' due to the dilution.

Thus, Chao's experiment also focuses on the second attribute of life: evolution.[1]

Bacteria on Earth can be shown to evolve in a laboratory as they are transferred from one culture to another for hundreds of generations and become better adapted to a given environment (substrate, temperature, etc.). Parameters such as density, growth rate, and lag time determine the increased adaptability of the microorganism. Within a month, a bacteria will be able to enhance its ability to metabolize in a given environment.

Chao's LR experiment can take advantage of this attribute by allowing for a large number of generations to be produced. If Martian microorganisms are subject to evolution, they will become better adapted to the environment given to them. Monitoring the release of labeled gas for each generation will be sufficient in detecting an evolutionary trend as the production of the gas should increase with time. This pattern is exclusively the realm of the living as far as we can tell; therefore, the chances of having a false positive using Chao's experiment are most likely null. If such an experiment landed on Mars and an increase in gas was detected with time, we would be more confident in our assertion that a process similar to life, and therefore life itself, has been found.

Lin Chao's thought experiment sheds some light on the difficulties scientists face when interpreting the Viking results as well as the many pitfalls inherent with experiments focusing exclusively on metabolic processes. Nowadays, space agencies shy

---

[1] Darwinian evolution is the process by which organisms change over time as a result of changes in heritable physical or behavioral traits. The changes which increase the chances of survival in a given environment will be more likely to pass on to the next generation. Unfortunately, Darwinian evolution is often misunderstood by the public, which sometimes confuses it for the defunct Lamarckism theory that supports the idea that an organism passes to its offspring the attributes that it has acquired throughout its life.

away from such experiments in their search for alien life, and as we saw, prefer to implement a broad strategy to characterize past and present habitable environments before moving on to the search for molecular biosignatures with missions such as the Perseverance and Rosalind Franklin rovers.

Living on the Edge

Now that we have roughly defined life and the attributes that characterise it, let us try to determine where such life could take hold within the environments present in our solar system. There is a catch here, though. By attempting to specify the chemical and physical properties of a habitable environment, we are hopelessly biased towards what we consider to be habitable environments here on Earth. To visualise what this represents, the habitable environment on our planet can be thought of as an enclosure surrounded by a fence of six physical and chemical extremes: pressure, high and low temperature, salinity, radiation, pH, and desiccation.

As you can see from Fig. 3.3, all known life on Earth resides inside this enclosure. With time, this enclosure gets bigger as we discover new organisms capable of surviving conditions, we previously thought were too extreme for life. The most remarkable and famous example of such discoveries occurred in 1977 when oceanographers stumbled upon hydrothermal vents deep within the Galapagos rift in the Pacific Ocean. These vents are fissures occurring on the oceanic floor from which geothermally heated water escape, providing a unique ecosystem of very complex life form totally independent from the sun's energy.

In the environments created by these vents, a diversity of life thrives undeterred by the superheated plumes (250 to 400 °C) saturated in toxic chemicals and heavy

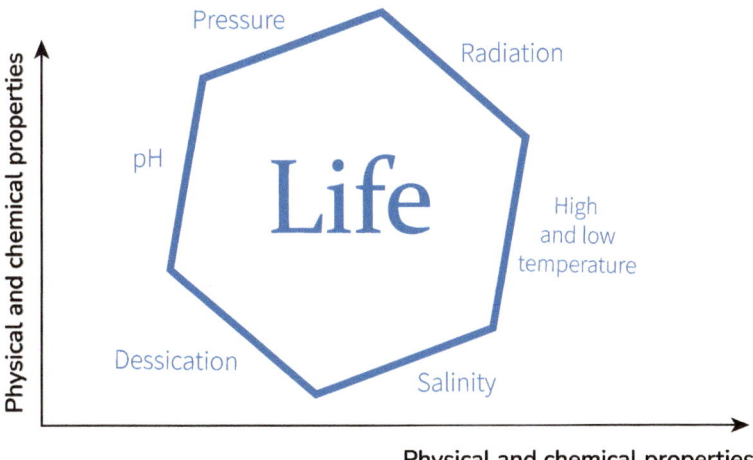

Fig. 3.3 The habitability of our planet is delimited by physical and chemical properties.

metals gushing out of chimney-like structures. Forming the base of the food chains found around the hydrothermal vents are microorganisms—bacteria and archaea— capable of surviving in absurdly hot temperatures. One of these exceptional organisms, Methanopyrus Kandleri, found on chimney walls, can withstand temperatures of up to 122 °C, a record no other organism has beaten yet. But what made these life forms genuinely alien to the biologists at the time was their capacity to harvest energy from the minerals and chemicals that belched out from these vents, a process known as chemosynthesis as opposed to photosynthesis. Within such an extreme environment, life has learned to transform hydrogen sulphide, which is a very poisonous and corrosive gas spewing out from the cracks, by causing a reaction called redox. What comes out from this reaction are new compounds such as sulphur and more importantly, energy. Hydrothermal vents represented a revolution in biology as, contrary to what was thought previously, life was found to thrive in the absence of energy-giving sunlight.

Indeed, forming the basis of a complex food chain, we have the rich microbial mats extracting chemical energy from the vents. Further up are clams, snails, mussels, and worms who graze upon these microorganisms or benefit from them through a symbiotic relationship. These then become prey for shrimps, anemones, and crabs, who in turn feed bigger crabs and fish as well as octopi. Within these rich ecosystems, referred to as hydrothermal vent communities, over 300 species of animals have been found, mostly living nowhere else on the planet. The largest animal discovered living around these vents are giant tubeworms that can grow up to 3 m long. These worms have no mouth or digestive system. Instead, they rely on a symbiotic relationship with bacteria that live inside them and use the hydrogen sulfide and other chemicals from the hydrothermal vents to produce food for the worm; the tubeworm gets its energy from the food that the bacteria produce while the bacteria get a comfy and safe place to live.

Once a vent stops being active, through earthquakes shifting the heat convection below, the food source disappears and the microorganisms perish. The rest of the fauna either migrate to a new vent if they can or die as well. Over a hundred life-bearing hydrothermal vents have since been discovered across the oceans with the most famous being found in the Mid-Atlantic ridge, the East Pacific rise, the Galapagos rift, the Guaymas basin, and the Logatchev hydrothermal field (Fig. 3.4).

Somewhat similar to deep-sea vents are cold seeps. These are areas on the seafloor where hydrogen sulfide, methane, and other hydrocarbon-rich fluid seepage occurs due to tectonic activity and get diffused by sediment, often over an area several hundred meters wide. Contrary to deep-sea vents though, the temperature of a cold seep is typically between 4 and 10 °C, hence the name. Microorganisms found in cold seeps once again find ways to extract energy through chemosynthesis.

Another seafloor environment where life challenges our presumptions are the hauntingly beautiful brine pools. In these pools, which range in size from a few hundred square meters to about 10 km², water has a salinity that is three to eight times greater than the water that surrounds them, and oxygen is rare or non-existent

Fig. 3.4 A biological community of mussels, shrimp, and limpets living at NW Eifuku seamount in the Marianas region. (Image courtesy of Submarine Ring of Fire 2014—Ironman, NOAA/ PMEL, NSF)

(anoxic). In fact, these bodies of water are so dense and heavy with salt that they lay on the seafloor like pools or small lakes.

Often labelled the pits of despair, animals such as crabs or hagfish who dare venture into these ultra-dense pools will experience toxic shock and quickly die. Yet, bacteria and archaea that can breathe methane—often present in abundance in brine—have evolved to survive in such extreme environments and form the basis of a chemosynthetic oasis where a community of highly specialised organisms thrive on the border of the toxic dead zone. The most important of these organisms are mussels that symbiotically partner with the methane-breathing bacteria and gather in large colonies on the edge of the halocline, the visible boundary between density layers. These large mussel colonies then support more complex animals such as shrimps, crabs, and hagfish.

It seems though, that despite the seemingly inhospitable conditions found in brine lakes, life finds a way. In 2010, scientists were stunned to discover multicellular animals capable of living in the Atalante Basin, an anoxic brine lake lying 3 km at the bottom of the Mediterranean Sea, where only unicellular organisms like bacteria and archaea were known to survive until then. Indeed, a team of Italian and Danish researchers sampled the Atalante brine and found three species of multicellular animals, loriciferans, that apparently spend their entire lives in oxygen-starved waters. Loriciferans are microscopic marine sediment-dwelling animals which were only discovered in 1983, of which the species discovered living within the Atalante

Basin are the first examples of animals capable of living without oxygen, raising the possibility for complex life to exist in similar environments elsewhere in our solar system.

Scientists have discovered deep-sea brines in only three bodies of water: the Gulf of Mexico, the Mediterranean Sea, and the Red Sea. Due to the ecosystem that surrounds them, brine pools are areas of high diversity separated by the vast barren plains of the sea floor.

It is worth mentioning here that even in the saltiest brine lake ever found, a small land-based brine pond nestled within a dry valley in Antarctica and named the Don Juan Pond, algal matts and muds contain abundant nutrients and a rich and varied population of microorganisms thanks to, or in spite of, its syrupy brine rich in calcium chloride allowing water to remain liquid at −50 °C.

Less famous, but still striking nonetheless, is the discovery of extremophiles living happily within the most extensive natural asphalt lake in the world, Pitch Lake on the Caribbean Island of Trinidad. Oil samples from the lake contained bubbles of entrapped water droplets of 1 to 3 μl (microliters) where unique bacteria can be found.

Adventurous microorganisms have also been found deep in Earth's outer crust, the layer just before the mantle, forming unique biospheres despite the absence of light and oxygen. Biologists are even considering that life might also be present in Earth's mantle which starts 6 km under the seafloor on average, and excavations done in 2015 in the Atlantis massif, a rocky region situated on the western side of the Mid-Atlantic ridge, suggest that this might indeed be the case. Future expeditions to drill into the mantle itself—a first—are planned for 2030 at the latest; these have the potential to uncover new ecosystems. As such, life's ability to thrive in Earth's outer crust, and possibly its mantle, might open the door to a potential ecosystem in the context of Mars exploration. That both simple and complex lifeforms manage to survive in these habitats would have seemed unthinkable a few decades ago, and yet, we are detecting them.

Among all the extreme environments on Earth, the ones discovered in Antarctica appear, as per our understanding, to bear the closest resemblance to the subsurface oceans of Europa and Enceladus. Already, astrobiologists are investigating the dry valleys of Antarctica as a primary destination for the search of analogues to past Martian environments. There, on windswept mountain slopes where extreme cold and aridity create an inhospitable environment for most organisms, minute lifeforms such as algae, lichens, and bacteria thrive within the rocks. Biologists refer to these microscopic organisms as cryptoendoliths (from the Greek, crypto = hidden, endo = in, lith = rock). There is speculation that if Mars once harbored life, its final remnants could resemble cryptoendoliths surviving in such unexpected sanctuaries.

As can be seen from Fig. 3.5, thanks to the Curiosity and Perseverance missions, we have discovered habitable zones on early Mars which overlap habitable zones found on Earth, raising the prospects of the possibility that Martian life might have thrived in the past.

However, within the context of this book, our fascination lies in what resides beneath the expansive ice sheets of Antarctica. Researchers were able to access a

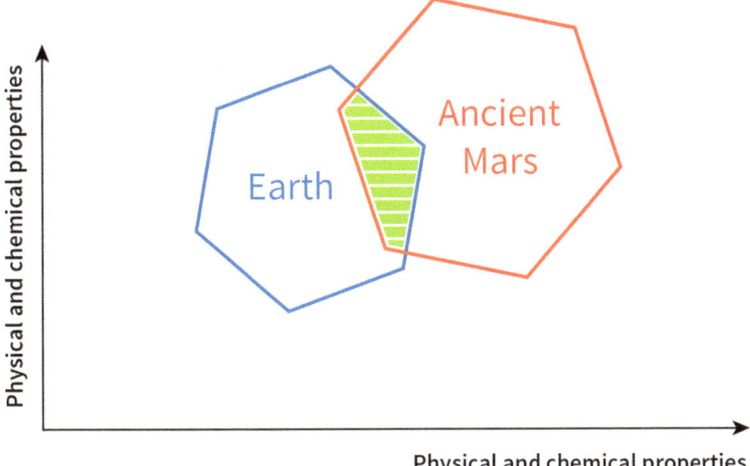

Fig. 3.5 The habitability of extreme environments discovered within our solar system can exhibit similarities to those found on Earth. In this instance, we draw a comparison between early Mars and the environments present in Antarctica's dry valleys.

subterranean river 500 m below the ice located hundreds of kilometers away from the edge of the Ross Ice Shelf. Scientists have known about a network of freshwater lakes and rivers flowing beneath the Antarctic ice sheets for a while, but they have not yet been able to study them directly. There, they found a swarm of crustaceans around 5 mm in size, pointing to a rich ecosystem far removed from open waters.

One of the most compelling finds was made in 2021 when sponges and unidentified organisms including barnacles and stalked animals were found on a rock 900 m below the Filchner-Ronne Ice Shelf, in waters lying 260 km away from the ocean. The study of local currents suggests that this community is somewhere between 625 and 1500 km in the direction of water flow from the nearest region of photosynthesis. This is remarkable as according to the established theory of sub-ice shelf biogeography, it is believed that biodiversity declines as one moves away from the ice shelf front along a nutrient gradient, although this is also dependant on ocean circulation, with richer and more diverse ecosystems associated with nutrient-rich inflowing currents and more impoverished seafloor coinciding with nutrient-poor outflow. Nevertheless, the existence of sponges and stalked animals found under the Filchner-Ronne Ice Shelf requires biologists to rethink the ability for life to survive in even the bleakest of places.

As we continue to explore the region, over a dozen deep drills have been made across Antarctic ice sheets throughout the last decades. The recent drill at Lake Whillans is the furthest "inland" from the ice shelf front where life has been observed. However, to better grasp the challenges faced by a hypothetical life in subsurface oceans like those on Europa or Enceladus, we need to examine subglacial lakes. There are around 400 known subglacial lakes under the surface of

Antarctica. These lakes are completely secluded from sunlight and have minimal interaction with the surrounding seas, rendering each environment unique.

After more than 10 years of planning, the British Antarctic Survey attempted, in 2012, to explore the subglacial Lake Ellsworth lying 3.4 km beneath the West Antarctic ice sheet only to abandon halfway due to unsurmountable difficulties. A year later, the first successful retrieval of a clean sample from an Antarctic subglacial lake took place at Lake Whillans located 800 m under the surface of a glacier on the other side of West Antarctica at 700 km from the nearest open water. The borehole, no bigger than 60 cm in diameter, allowed samples of freshwater as well as sediments at the base of the lake (around 2 m in depth) to be retrieved, after being trapped under the ice for more than 120,000 years. The samples contained almost 4000 species of single-celled organisms thriving on the sediment on the lakebed despite the water being −0.49 °C.

Yet, no review of subglacial lakes in Antarctica would be complete without mentioning the holy grail of Antarctic exploration: Lake Vostok. Everything about Lake Vostok fascinates. First suggested in the early 1960s, Lake Vostok holds a volume of water bigger than Lake Michigan and is the largest of Antarctica's subglacial lakes with an area of 12,500 km^2 It is situated at roughly 4 km under the surface of the ice and 500 m below sea level, with a maximum depth of up to ~900 m making it amongst the deepest lakes in the world.

Its confirmation in 1993 has generated a great deal of excitement among scientists as it is thought to have been isolated from the outside world for over 15 million years when it was first covered with ice. This isolation has made Lake Vostok one of the most pristine and mysterious lakes on Earth and might contain valuable information about the history of life. By pure coincidence, of all the places the Soviet Union had decided to build their Antarctic research station in the late 1950s to conduct atmospheric, geological, biological, and climate research, they selected a site right on top of the yet-to-be-known lake in this remote location in East Antarctica; Vostok means East in Russian.

Thus, the Russians were best placed to attempt deep drillings within the ice under the Vostok station. Alas, in their haste to be the first to reach the lake, they employed crude technology that required freon and kerosene drilling fluid, which are chemically and microbiologically dirty. In 2012, when the Russians finally reached Vostok lake after drilling through 3769 m of ice, they allowed lake water to gush up the borehole to 363 m and freeze. This water got mixed with the freon and kerosene surrounding the drill. Not only was there a risk of contamination within the lake itself, but the 32-m ice core of frozen lake water retrieved for analysis had been compromised, shedding doubts on any biological results put forward.

Another hole was drilled 3 years later, with lake water rising through the borehole up to 70 m before it froze. Within 4 days, this icy 'cork' was drilled into, and frozen lake water was collected into a sterile receptacle before water pushed itself through the borehole once again.

In both cases, the water samples were rather clean but still contained numerous microdroplets of drilling fluid, giving the ice a 'milky' appearance. Nevertheless, the samples were analyzed, unveiling a thriving microbial community. DNA sequencing of ice cores has identified hundreds of bacterial species, as well as archaea and fungi. The most intriguing discoveries include genetic material potentially associated with multicellular organisms. These include microscopic crustaceans, water fleas, a mollusk, and even a sea anemone. However, the possibility of contamination from drilling activities or past interactions with the ocean cannot be completely discounted. Due to the possibility that the lake has been sealed off for millions of years, Vostok lifeforms could hint at some of the evolutionary pressures that might also exist under Europa and Enceladus.

By studying extreme environments on Earth, it becomes evident that wherever liquid water, rocky material, and a heat source converge, life can thrive. This understanding has expanded the boundaries of habitable environments on our planet, as life continues to get discovered in diverse locations ranging from the upper atmosphere to the depths of the Earth's crust. It suggests that life has remarkable adaptability to a wide range of conditions, raising the possibility of it being present in similar environments elsewhere in our solar system.

As we saw earlier, when we try to characterize the habitability of an environment such as the Martian surface or Enceladus' subsurface ocean, we are in fact searching for an overlap with known habitable areas here on Earth. If such overlap exists, then life might have been present there and thrived, thus providing a strong case to investigate it. The good news is that we are continuously discovering extremophiles in places where we thought life couldn't take hold, thus increasing the range where an overlap might be possible with alien environments.

Let us now investigate the key requirements for making an area habitable.

The Three Ingredients for Life

Embedded at the very heart of the habitability concept lies the conditions required for life to emerge. Although bacteria and archaea can survive in extreme environments such as Pitch Lake, these microorganisms didn't live inside the hydrocarbons themselves but rather in tiny bubbles of water.

As such, biologists often list three conditions essential for life: liquid water, an energy source, and access to nutrients and organic compounds (the building blocks of life). For complex life to emerge, another condition is required: time. Let us review each of these conditions within the context of subsurface oceans.

Water is vital for all known forms of life, despite not providing food energy, or organic micronutrients. It is essential though to make a distinction between water in its liquid state, which is also called water, and water its solid state, known as ice, as well as its gas state, known as vapor. Whenever needed, I will clarify what state the water will be as to prevent any confusion.

To this day, we have yet to find a living organism that has been proven to survive without water in its liquid state. There are of course extremophiles that can survive—just barely—in nearly waterless environments such as in the arid valleys of Antarctica but eventually, even these extraordinary microorganisms will require a minimum of liquid water to function, sometimes having to wait thousands of years to do so.

It is therefore safe to assume that without liquid water, life on Earth as we know it wouldn't exist. This is mainly thanks to the fact that water molecules have this unique ability to act as a medium for organic compounds, mixing them in the process. As a matter of fact, liquid water is known as the perfect universal solvent, as it can dissolve more substances than any other liquid. This characteristic is due to its simple polar arrangement of hydrogen (positive electrical charge) on one side and oxygen (negative electrical charge) on the other, which can disrupt the attractive forces that usually keep other molecules together.

One of the other essential characteristics of water is that, because it is a polar substance, polar compounds are more soluble in it than non-polar ones (e. g., long alkanes chains). This means that compounds with hydrophobic tails and hydrophilic heads can 'clump' together to form micelles, aggregates of molecules in water that are typically spherical in shape; this is how soap works as the tails dissolve in the fat and the heads in the water. The natural appearance of micelles in water is actually thought to be one of the mechanisms for the early formation of cell membranes, key to carbon-based life. In addition, being able to dissolve most substances means that wherever it goes, liquid water will carry with it valuable nutrients and minerals, making it an ideal method of transportation within the medium into which a living organism appears but also within the organism itself as it evolves into greater size and complexity. Liquid water also acts as a physical barrier, shielding potential life-bearing habitats from harmful solar radiations, highly charged particles and other nasty stuff coming from deep space.

When considering extraterrestrial life-forms, some scientists consider moving away from the water-centric approach by suggesting other forms of solvents for life. Ammonia, NH_3, springs to mind. Liquid ammonia can become a medium for many compounds. Its downside are the extremely cold temperatures at which it only exists as a liquid, between -33 to -78 °C, where it then solidifies into a mass of white crystals. This temperature range makes it difficult for metabolic reactions to occur.

Other solvents proposed are liquid hydrocarbons, such as methane or ethane, which form lakes on the surface of Titan, Saturn's giant moon. There are several factors that make liquid hydrocarbons a potential solvent for life. First, liquid hydrocarbons are polar molecules, just like water. This means that they can form hydrogen bonds, which are important for the structure and function of biological molecules. Second, liquid hydrocarbons are good solvents for organic molecules, which are the building blocks of life. Thus, it has been postulated that a silica-based life could also grow in a liquid methane habitat, since methane is a suitable solvent for silanes (SiH_4) and polysilanes (compounds of multiple silanes), thereby mimicking the liquid water and organic chemistry association found on Earth. However, there are also some challenges to the idea of life in liquid hydrocarbons: hydrocarbons are not as good of solvents as water, which will limit the complexity of structures that can be built, thus also reduce the potential for life to arise. Also, liquid hydrocarbons are more volatile than water, which means that they evaporate more easily. This could make it difficult for life to exist in liquid hydrocarbons on planets or moons with a thin atmosphere. Despite these challenges, the possibility of life in liquid hydrocarbons remains an intriguing possibility, as in theory, life forms could exist in such habitats by using hydrogen and acetylene as an energy source.

The only certainty we have for now though is that without liquid water, carbon-based life forms cannot exist, which is why our strategies for the search for life in the solar system involves searching for liquid water, either on Mars or within the depths of icy moons.

The next essential condition for life to emerge is energy, which can be roughly defined as a property that can be stored or transferred to an object that will give it heat or allow it to do work. Energy can take many forms, such as kinetic energy (motion), electrical energy, radiant energy (light), thermal energy (heat), nuclear energy, chemical energy, gravitational potential energy, etc. Without energy, molecules and atoms become inert (cold), preventing any reactions whatsoever. Although the requirement for liquid water as a critical condition for life might be too Earth-centric for some, the need for an energy source is universally accepted.

An energy source is required to keep the temperatures of a habitat within the range in which life operates: from −25 to 121 °C. Note that salts can lower the freezing point of water and life has been found to manufacture antifreeze compounds to survive in even colder environments. Within the context of icy moons and dwarf planets, we can imagine hydrothermal vents, occurring at the intersection between a rocky mantle and an icy crust, radiating enough heat to melt small pockets of frozen water around them. But it is doubtful that life could appear in such occasional environments. Better to consider the large bodies of liquid water such as the liquid mantles found under Europa or Enceladus or the hypothetical broad sea under Pluto, where the chances for life to appear are better. Yet, because such bodies represent a considerable volume of water, only a few processes can generate enough energy to melt them: primordial heat, radiogenic heat, and tidal heating.

Note that asteroids and comet strikes can also provide heat, although only localized heat. In addition, while collisions with planetary bodies were very common in the early history of our solar system, this form of heating isn't pertinent to bodies of liquid water locked under thick layers of ice.

Let us cover the three types of energy listed in the previous paragraph. Primordial heat is the heat that was trapped inside a planetary body as it was forming in the early solar system. It consists of two forms of heating. The first is accretion heat, which is generated by the conversion of the kinetic energy of impacting bodies to thermal energy. Despite occurring billions of years ago, we are still benefiting today from the energy unleashed by a massive collision between Earth and a protoplanet, which resulted in the birth of our Moon. The second form of heat is referred to as gravitational release, which occurs when a planetary body has accumulated enough mass and heat to go through differentiation: the separation of different constituents of a planetary body. As denser material moves towards the center, such as iron, friction is created between the moving masses and converted to heat. (Phase transitions, such as when water vapor turns to rain, can release a lot of latent heat as well. An example is the case of Saturn, which is found to be warmer than it should be from solar gain alone, and this is thought to be due to 'helium rain' falling towards the center, releasing heat as it goes.)

Accretion heat generates a positive feedback loop, as the melting of icy and rocky material add further energy into the system. Once differentiation is complete, this form of heating ceases to be generated. However, the energy it has released will be stored within the interior of a planetary body for a very long time. On Earth, primordial heat still represents half of the total energy released by our planet today even though accretion heat and gravitational release first occurred 4.5 billion years ago. But the earth is relatively large for a rocky object. Bodies of smaller densities and volumes, such as icy moons, will not be able to hold onto such energy for as long and will dissipate the heat more quickly. The problem with smaller bodies is that of absolute size, as the surface area goes up with the square of the radius, but the volume goes up with the cube. Therefore, smaller objects cool faster.

Another type of energy is radiogenic heat, which is generated from the decay of the nuclei in radioactive isotopes (radioisotopes), unstable elements that dissipate their excess energy by emitting radiation in the form of alpha, beta, and gamma rays. These isotopes are present in the mantle and crust of planetary bodies (lithosphere) and warm these parts uniformly. The thicker the mantle and crust are, the more it warms up. As an example, Earth's radiogenic heat is responsible for the other half of the total heat produced by our planet. Mercury, on the other hand, has smaller layers of crust and mantle, which in turn reduces the amount of energy it receives from radiogenic heating. However, due to its small size, it should take less time to heat it up. Regardless, due to the relatively small sizes of their lithosphere, some moons have little to no radiogenic heating. Note that this form of heating can be long-lasting due to the presence of long-lived radioisotopes whose half-lives are counted in hundreds of millions or billions of years.

Tidal heating, introduced in chapter one, has its roots in the gravitational pull of one or more planetary bodies on a satellite. The variation in heating depends on the severity of the gravitational pull as well as the internal composition of the satellite (icy or rocky). Like most things in life, moderation is best. Too much and you end up like fiery Io scarred by the constant volcanic activity, too little and you end up like a frozen snowball.

Tidal heating has two significant advantages over primordial and radiogenic heat. First, it can last almost indefinitely once the orbits of celestial bodies are locked into place. Io, Europa, and Ganymede have been subjected at varying degrees to tidal heating for billions of years. This is more than enough time for things to get interesting as far as life is concerned. Not every orbit is stable, though, and in that case, tidal heating can occur for a shorter period—but still within hundreds of millions of years. The Saturnian satellite system comes to mind as the small icy moons seem to drop in and out of resonance with each other throughout their formation.

The second advantage of tidal heating is that, depending on the circumstances, the energy can be focused on a specific region inside the planetary body. The heating then becomes localized. Therefore, an icy moon might not receive enough tidal heat to melt its entire ice mantle, but it could still host a body of liquid water confined to a particular area within it.

We shall revisit these three forms of energy as we explore one by one the confirmed ocean worlds or the planetary bodies which we suspect might host a subsurface ocean. Let us now move to the last condition we believe life requires to get started: the presence of organic chemistry.

Organic Chemistry for Dummies

The assembly of complex organic compounds forms the foundation of all life on Earth. Yet, this can only occur if two prerequisites are met: simple organic compounds need to be present to assemble the molecular structures that life requires, and an energy source needs to be present to drive metabolic reactions; on Earth, the most abundant sources of energy are solar energy and chemical energy from redox reactions.

Organic chemistry is the study of these two prerequisites.

The good news with the first prerequisite is that a rich variety of simple organic compounds can be found in the interstellar medium, planetary atmospheres and surfaces, comets, asteroids and meteorites, and interplanetary dust particles. These include carboxylic acids, sulfonic acids, sugars, urea, aliphatic and aromatic hydrocarbons, ketones, ammonia, alcohols, purines, and pyrimidines just to name a few. The building blocks of life lie everywhere we look. More than 70 different amino acids have been identified in meteorites alone. In September 2017, ESA announced, after studying at length the data returned by the Rosetta mission as it investigated Comet 67P Churyumov-Gerasimenko, that organic matter made up 40% by mass of its nucleus and proposed as its origin interstellar space instead of within our solar system. As such, comets like 67P Churyumov-Gerasimenko must have delivered vast amounts of organic matter to Earth through the early bombardment phases of our solar system, and they still deliver small amounts today. When astronomers look further away, organic molecules have been discovered not just in interstellar clouds from the Solar neighbourhood, but also throughout the Milky-Way, as well as in nearby galaxies, or some of the most distant quasars.

Furthermore, organic compounds—and therefore life—are formed mostly by using six essential elements only: carbon, hydrogen, nitrogen, oxygen, phosphorus, and sulfur. An easy catchword has been created to label them: CHNOPS (or SPONCH). The four major classes of organic molecules: carbohydrates, lipids, proteins, and nucleic acids are built entirely from the CHNOPS elements.

The presence of organic compounds throughout the universe is not entirely surprising given that these six elements make up the majority of matter in our universe (see Table 3.3): hydrogen, formed at the Big Bang, makes up three-quarters of all the mass in our observable universe, followed by oxygen (third most common element), carbon (4th), nitrogen (6th), sulfur (10th), and not seen on this table, phosphorus (19th). Life didn't rise from rare elements that were difficult to combine. Instead, early life used stuff that is easy to assemble and present everywhere.

Table 3.3 Chemical elements ranked by their relative abundance in space. Five of the six essential elements for life to come into existence are part of the ten most abundant chemical elements in space. Phosphorus (P) is the 19th most abundant element.

Abundance of chemical elements in our Universe					
Rank	Symbol	Elements	Universe	Sun	Earth
1	H	Hydrogen	92%	94%	0.20%
2	He	Helium	7.10%	6%	0.00%
3	O	Oxygen	0.10%	0.06%	48.80%
4	C	Carbon	0.06%	0.04%	0.02%
5	Ne	Neon	0.01%	0.00%	0.00%
6	N	Nitrogen	0.02%	0.01%	0.00%
7	Si	Silicon	0.01%	0.01%	13.80%
8	Mg	Magnesium	0.01%	0.00%	16.50%
9	Fe	Iron	0.00%	0.00%	14.30%
10	S	Sulfur	0.00%	0.00%	3.70%

Let us review these six elements and see what molecules they form as we combine them. We will start with the carbon atom, the heart of organic chemistry. Carbon is a remarkable element. It bears different forms in nature, from graphite, the softest, to diamond, one of the hardest known substances. Being tetravalent, each atom of carbon has four electrons capable of forming multiple bonds. This and the atom's small size makes it one of the most versatile elements, which explains why the Beilstein database, the most extensive database in the field of organic chemistry, contains almost ten million organic compounds identified so far.

By adding hydrogen atoms to a carbon atom, a hydrocarbon is formed. This is the simplest of the organic compounds with the smallest hydrocarbon being methane (CH_4) composed of one carbon atom and four hydrogen atoms. Ethane (C_2H_6) has two carbon atoms bonded to each other and surrounded by six hydrogen atoms. Propane (C_3H_8) has three carbon atoms bonded to each other and surrounded by eight hydrogen atoms, while butane (C_4H_{10}) has four carbon atoms surrounded by ten hydrogen atoms, and so forth. As we add more carbon and hydrogen atoms to the hydrocarbon, the chain gets longer. We call these polymers. When the chain of hydrocarbon is between five to nine carbon atoms long, it is named gasoline. At about a dozen carbon atoms long, it is diesel, and around 20 carbon atoms long, it becomes motor oil. Carbon is unique in its ability to form polymers at the relatively low temperatures found on Earth. This makes it an ideal element for building molecular structures, as it can form large and complex molecules without the need for high temperatures or pressures.

If we insert oxygen between carbon and hydrogen, methane becomes methanol ($CH_4 \rightarrow CH_3OH$), ethane becomes ethanol (C_2H_5OH), propane becomes propanol (C_3H_7OH), and so forth. These are all known as alcohols which are found across space. For example, methanol has been detected in molecular clouds, protostellar regions, and regions around dying stars.

However, you can also make another class of organic compound using these oxygen, hydrogen, and carbon. If you swap two hydrogen atoms from the alcohol chain with a single oxygen atom, you form carboxylic acids, commonly known as organic acids. These are called acids because the hydrogen bonded onto the oxygen will tend to come off easily, making it quick to react with other elements—more precisely, the proton from the hydrogen atom will leave the carboxylic acid while the electron of the same atom will stay. The smallest of these acids is formed when a hydrogen atom in methanol is replaced by an oxygen atom (CH_3OH becomes HCOOH), forming formic acid found in the venom of some ant species. Acetic acid (CH_3COOH), also known as vinegar, is derived from ethanol and is created when an oxygen atom replaces another hydrogen atom. Butyric acid (C_3H_7COOH) appears when butter goes rancid. Citric acid ($C_6H_8O_7$) is present in citrus fruits.

Now things get interesting. If you combine alcohols and organic acids, casting off a molecule of water in the process, a new class of organic compound is formed, the esters. Esters are the basis for fragrances. For example, ethyl butyrate has pineapple flavor. They also form one of the main classes of lipids (found in animal fats and vegetable oils). Alcohols, carboxylic acids, and esters have been found within large interstellar clouds.

Apart from alcohols and organic acids, another class of compounds use carbon, oxygen, and hydrogen. These are known as the carbohydrates. The name carbohydrate indicates the presence of carbon and water. Carbohydrates are used to form cellulose, starch, sugars, and glycogen.

It is surprising how much you can do with the first, third, and fourth most abundant element in our universe. There is more though. If you take aliphatic compounds, which is a type of long hydrocarbon chain, and add only a handful of oxygen atoms in the form of carboxylic acid, you can create fatty acids useful for energy storage and cell membranes. Thus, hydrocarbons, organic acids, fatty acids, esters, lipids, carbohydrates, and alcohols are formed with only three elements. Not bad. But we still have a few more elements to go through.

Let us now add the next element to this already remarkable mix: nitrogen (N). Combine hydrogen to nitrogen, and ammonia is formed (NH_3). This is a common volatile compound found in the outer solar system. If you add acetic acid to ammonia, glycine, the simplest of the amino acids, is formed, opening the door to more complex amino acids. In turn, amino acids are required for the formation of proteins. Proteins are useful as they allow life forms to build structures in an organism: keratin for hair, collagen for skin, myosin for muscles, etc. Throw in the next essential element, sulfur (S), and two main sulfur-containing amino acids are possible, cysteine and methionine, which play an important role in protein structure, enzymatic reactions, and other biological functions.

Finally, phosphorus (P), the last essential element to be added to this life-bearing mix, will bring about the nucleic acids. We have all heard of the two main types of nucleic acids: deoxyribonucleic acid is a long molecule known as DNA, and ribonucleic acid know as RNA. Incredibly, the instructions for all life on Earth is written with only six elements: C, H, N, O, P, and S (Fig. 3.6).

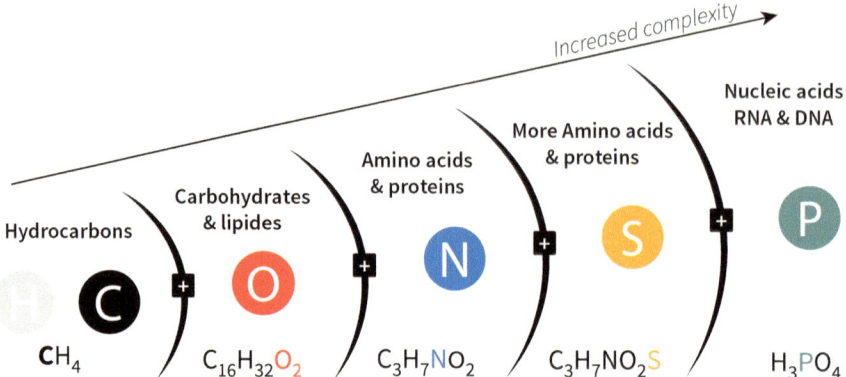

Fig. 3.6 The six elements and how they contribute to the families of organic compounds.

We are lucky. The staggering complexity of life on our planet is mainly derived from only six elements out of 92 that occur naturally. Such a small set of essential elements for life makes the quest for habitable environments significantly easier. For example, all these elements have been detected within the plumes of Enceladus.

Food Sources

Now that we have been introduced to the building blocks of life and how these are formed with just a handful of elements, let us see how life sustains itself once it gets started. On Earth, light and inorganic chemicals capable of undergoing a chemical reaction provide the two fundamental energy sources for life; once organic matter is produced, it becomes the third source of energy which we depend upon.

Regardless of their origin though, the electron is at the heart of all energy sources. Within a type of chemical reaction called reduction-oxidation reaction, or redox reaction for short, the negatively charged subatomic particle is transferred from one atom to another, and it is this transfer of electrons that unleashes energy, making it an energetically favorable chemical reaction. Some of the most common redox reactions found in nature are photosynthesis, hydrogen oxidation, methanogenesis, sulfur reduction and oxidation, iron reduction (rust), denitrification, and aerobic respiration. Note that a lot of energetically favorable acid-base reactions are also present in many of these processes. Redox reactions have also been found to use uranium, copper, arsenic selenium, and lead, although these are rarer.

Within some extreme environments on Earth, life has learned, among other things, to create a redox reaction by transforming hydrogen sulfide into sulfur, thus creating new compounds and energy. It is worth remembering that before hydrothermal vents were found at the bottom of the eastern Pacific Ocean in 1977, life was known to be entirely dependent on the energy of our Sun enabled by the photosynthesis process. Places where the sunlight was non-existent, such as deep within Earth's crust or on a distant planetary object, was considered off limits to life. Instead, the discovery of deep-sea vents turned the idea of habitability on its head as biologists could now consider sunlight-deprived habitats to be potentially conducive to life, not inhibiting it.

And so, if a heat source is present, and liquid water is in contact with rocky material, life seems to gain a foothold. Therefore, in addition to liquid water, organic compounds and heat, life needs minerals from rocks. For that reason, it is essential to understand where the rocky material is located within a planetary object. In big moons massive enough to have undergone differentiation such as Europa or Ganymede, rocks tend to be concentrated within rocky mantles also known as a silicate mantle. Within smaller entities like the dwarf planet Ceres, even if complete differentiation hasn't occurred, certain areas could potentially contain a blend of rocks, ice, and even liquid water.

This simple, yet crucial fact carries important implications for the potential existence of life within subsurface oceans. If a body of water lying within an icy moon is not in direct contact with the rocky mantle or significant amounts of rocky material, the minerals and chemical reactions required to fuel life will not be present. Therefore, understanding the internal structure of an ocean world becomes essential in evaluating its potential for habitability.

Ironically, there can also be too much of a good thing as some icy moons contain so much liquid water that the intense weight at the base of the water mantle forces the liquid water to change into a high-pressure crystalline form of ice, and will consequently, seal off a subsurface ocean from the rocky mantle below. Essentially, the subsurface liquid water ocean will be enclosed between two layers of ice: an icy crust above and an icy mantle below. Lacking any evident mineral source to sustain potential life within these enclosed global oceans, the establishment of life would face significant challenges.

Or maybe not. Recent studies seem to indicate that this might not be entirely the case, as we shall see in chapter four where we investigate Ganymede's massive ocean mantle in detail.

Europa on the other hand might have a relatively thin layer of water mantle (200 km) resting on top of the moon's rocky mantle. Despite the considerable pressures experienced at the lower part of this liquid mantle (as a reminder, Earth's average ocean depth is only about 3.7 km), they fall short of creating a solid layer of ice at the bottom. In this context, the surface areas where rock is being exposed to liquid water are extensive. In addition to the vast seabed, many cracks and fissures will most likely be present on the oceanic floor, allowing water to penetrate within the rocky mantle.

Relatively warm fluids will flow within the rock fractures, and given the right conditions, could bring about serpentinization, a low-temperature process that alters minerals such as olivine made of magnesium and iron. On Earth, when olivine is exposed to seawater, mineral components in the olivine, such as forsterite and fayalite, undergo an exothermic reaction that produces hydrogen gas and heat. The hydrogen atoms can be used as fuel for life. That is why scientists got so excited when they found hydrogen in Enceladus' plumes. It appears that if we were to introduce selected microorganisms from Earth into a subsurface ocean where serpentinization takes place, they would likely survive.

Even so, concluding that a subsurface ocean possesses the necessary conditions conducive to life such as heat, minerals, organic compounds, and liquid water is very different than claiming that it can arise there in the first place. What life requires to come into existence is still a matter of great debate within the biological sciences, and even though robust theories have been proposed, no consensus has been found yet. Some scientists strongly back the deep-sea hydrothermal vents, or where serpentinization occurs, as the ideal habitat for life to appear, whereas others suggest that pools of freshwater or hydrothermal volcanic fields located on Earth's surface (such as geysers and hot springs in Iceland and Yellowstone National Park) are more likely habitats.

For example, a paper entitled "Can Life Begin on Enceladus? A Perspective from Hydrothermal Chemistry," published in 2016 by David Deamer and Bruce Damer, both researchers in biochemistry and biomolecular engineering at the University of Santa Cruz, or by the same authors, 'The Hot Spring Hypothesis for an Origin of Life,' published in 2020, suggests that life may require cycles of hydration and dehydration in environments similar to hydrothermal fields located on the surface of a landmass where an atmosphere is present. If this proves to be correct, Enceladus or Europa would be habitable but lifeless. Interestingly, the conditions outlined by Deamer and Damer appear to have existed on early Mars, evidenced by the discovery of remnants of ancient shallow oceans and previous hydrothermal activity.

Given the ongoing uncertainty regarding the origin of life on Earth, three scenarios come into consideration:

- Life on Earth started from hydrothermal vents at the bottom of the oceans, in which case, we might find life in the subsurface oceans which supports such environment.
- Life on Earth started from hydrothermal fields on the surface of landmasses, in which case, we might not find life in subsurface oceans.
- Panspermia is common in our solar system, in which case, whatever life we find in the subsurface oceans of Europa or Enceladus might be similar to life on Earth with RNA or DNA serving as biological information systems.

As life is a complex unsolved puzzle, it might be that all three of these scenarios prove to be correct at the same time; life could arise from hydrothermal vents that are situated at the bottom of the ocean as well as on the surface of landmasses and later get reshaped by cross contamination. With this in mind, and given the current uncertainties, the view taken by this book is that if a subsurface ocean proves to be habitable, then life might also have arisen from it. The next question we are faced with is how long does it take for life to appear?

Time and the Complexity of Life

We have seen throughout this chapter that life requires liquid water, organic compounds, minerals from rocky matter and energy sources to generate heat and provide nutrients.

Once these conditions are met, how long does it take for life to get started? Ten million years? A hundred? A billion? This is a hard question to answer given our difficulties in determining exactly when life on Earth arose in the first place. Scientists have nevertheless been delving deep into the blurry past of our genesis, where much uncertainty remains, and a clearer yet still fuzzy picture has started to emerge following recent discoveries.

As far as we know, it goes something like this. After an intense period of bombardment by asteroids and other space rocks, Earth formed around 4.6 billion years ago with the rest of the solar system, thus starting the Hadean era, the first geological era of our planet, which lasted a little over 500 million years. Earth was still very hot as it had absorbed enormous energy from accretion and the abundance of short-lived radioactive elements. The surface was mainly composed of molten rocks, and an atmosphere was slowly starting to appear due to the existence of volatiles. During the Hadean, our planet would be pummelled occasionally with rocks from space, but nothing compared to the sheer intensity and scale as the initial accretion period. That is, except for two catastrophic events: the formation of our Moon and the Late Heavy Bombardment phase.

The formation of our satellite is still poorly understood. Some theories put the event at just a few tens of millions of years after the formation of our planet, while others place it at 150 to 200 million years, still very early on in geological and life-forming timescales. Also, it was assumed that the Moon was formed due to an impact on Earth by a single object with a mass similar to Mars called Theia; however recent studies seem to refute this model and instead support the idea that dozens of lesser impacts (one to ten masses of the Moon) might have done the trick. Regardless which theory is right, either events would have turned a cooling Earth into a state of fiery chaos and wiped the slate clean.

There was also a period of Late Heavy Bombardment occurring at around 500 million years after Earth's formation 4.1 billion years ago, which lasted hundreds of millions of years, and as its name suggests, inflicted much devastation to our planet's young surface.

For that reason, illustrations of this era often depict Earth as a hellish place where bright lava flows spewing out on a dry and desolate rocky surface. Until recently, no one seriously considered that life could arise in these conditions. Previously, the first tangible signs of life found were embedded in ancient rocks in South Africa and Australia as microfossils showing cell-like structures. These rocks have been dated at 3.5 billion years old, in the Paleoarchean Era, well after the end of the Hadean Era.

Before this period, no signs of early life could be uncovered. Actually, no rocks dating from the Hadean Era could be found, making any claims about this tumultuous period challenging to validate. Nevertheless, it was understood that life couldn't possibly have formed during the violent Hadean Era and that once the dust had settled and the conditions were right, it still took hundreds of millions of years for life to arrive on the stage. Life was thought to be picky and slow. In other words, it was rare.

However, as is often the case in science, fresh discoveries have forced us to re-evaluate this traditional view. In comes the zircon, a tough mineral with a cool name. Preserved deep within sandstone rocks in western Australia, scientists have found tiny grains of zircon that have been dated as the oldest fragments on Earth ever to be discovered at 4.4 billion years old, providing a unique window into the Hadean Era before the Late Heavy Bombardment of 4.1 to 3.8 billion years ago. Given the inherent difficulties in interpreting scientific results without context due to our lack of knowledge of the Hadean Era, there has been much debate as to what these zircons can tell us. These crystals would not have been able to form if the Earth's surface was as hot as previously thought and some scientists have proposed that the elements trapped within the crystals reveal a more peaceful period, where a vast original ocean prevailed. Far from the apocalyptic vision we once had, our planet might have been cooled down earlier than initially suspected, making it a more hospitable environment for early life. Such hypothesis is called the Cool Early Earth and supports the notion that the Earth's surface was much cooler than previously thought in the first few hundred million years of its existence. Others have challenged this interpretation, and the debate as to what these oldest zircons can tell us still goes on.

Furthermore, younger zircons, found in western Australia and dated at 4.1 billion years old, contain graphite flecks that could have a biological origin. If this turns out to be true, this will reveal a more hospitable environment, although once again other non-biological interpretations have been put forward.

More promising are the tell-tale signs found in some of the oldest rock formations on Earth in Isua, Greenland. These have shown intriguing similarities with what appears to be fossilized stromatolites, bulbous formations of sedimentary grains, built layer upon layer by bacteria. We know stromatolites well as they can still be found today in extremely saline lagoons. Since these structures were the result of a community of microorganisms displaying some complexity, life would have needed to start much earlier. The Isua rocks are dated at 3.8 billion years old, which if the interpretation of stromatolites is correct, would push the emergence of life millions of years earlier. Recent studies have also proposed that tube-like structures like those found in hydrothermal vents might have been discovered in rocks that are 3.77 billion years old as well as in the Isua rocks.

The discovery of stromatolites and other evidence of early life on Earth has led some scientists to believe that life may have arisen very early in the history of the planet and while it might appear that life was anything but picky and slow, it is important not to get too carried away with these recent discoveries, as much is still contentious. Trying to deduce the tell-tale signs of life in material that have been subjected to billions of years of physical and chemical processes can be misleading.

Recently though, a novel approach on this topic has brought up new insights that might support the interpretations of life starting early. By analyzing DNA databases of thousands of modern species, a new line of investigation called "molecular clock analysis" can trace the earliest points at which specific sequences have been expressed by ancestral cells. Studies using this tool have already suggested that the first animals emerged 1.2 billion years ago, several hundred million years earlier than the oldest fossils found and that eukaryotes (cells containing a nucleus) could have made an appearance much earlier than expected as well at around 2.3 billion years. Some scientists are now using this tool to go even further back in time to estimate when life first appeared and preliminary results have put it at around 4 to 4.1 billion years ago, which seems to support the recent fossilized discoveries. Although there is still much to be confirmed, it does look that we are moving away from the idea that life was slow to get going. Still, time is crucial, and some astrobiologists list Enceladus below Mars, Titan, and Europa for the likeliest places where a second genesis might have occurred based on the belief that its habitable zone is young.

There is as well the intriguing idea presented earlier that life might not originate from our planet at all but was instead delivered by asteroids and other space rocks. Referred to as the Panspermia theory, this idea implies that we could be descendants of life forms that might have initially started in the ancient oceans of Mars, Venus or maybe even deep within Ceres' interior. Past impact regularly threw pieces of crusts from Mars and Ceres in space which could have carried the alien microrganisms to our planet.

Regardless of where life originated, the most conservative view within the scientific community is that it requires more than half a billion years to appear once the conditions are 'right', while the most optimistic view proposes that life needed less than 10 million years to get started. At this point in time, it is anyone's guess as to which side more accurately reflects the reality, and we could do worse than suggesting meeting halfway at 250 million years.

What we do know, though, is that complex life forms took a very long time to appear. Precise estimates will vary depending on sources; nevertheless, they all present the view that after single cells emerged within a liquid environment, it took a staggering more or less 2 billion years to evolve into complex cells (Eukaryotes) and roughly another half a billion years to organize themselves as multicellular organisms (e. g., plants or bugs). That life on Earth needed 2.5 billion years to create a modest worm wriggling in a sandy seafloor suggest the need for geological time spans to form complex life forms—the presence of an oxygenated environment seems to have helped as well. Yet, the dangers of focusing on one sample only, Earth, in extrapolating life's capabilities are evident. It is possible that the conditions for life within subsurface oceans might be entirely absent due to a set of circumstances that remain currently unknown to us. The contrary might be true as well and complex fish-like alien lifeforms could be thriving in many oceans, although it might be that not enough energy is available in subsurface oceans to sustain the energy demands of complex creatures, no matter how romantic the concept might seem.

Now, let us envision a scenario where life does indeed originate in a habitable zone within an ocean world. In what manner might it manifest itself? Would it stay microbial or become multicellular? What body plan would it evolve into? In other words, what would they look like?

No one has a clear answer to such questions, although we can make educated guesses from the study of lifeforms that have arisen on Earth. For a start, let us not forget that the chemistry of life is an aquatic chemistry. Every living being can get by on land only by carrying a huge amount of salt water around with them. So, it is reasonable to suggest given the liquid environments that are central to this book, that life within these environments will probably share similarities with life that arose on Earth.

The primary classification of life on Earth comprises of three domains: two types of prokaryotes (organisms that do not have a nucleus or organelles) which are bacteria and archaea, and eukaryotes (organisms with complex cells containing a nucleus and which all complex life derives from). While we cannot be certain, it is highly probable that prokaryotes were the first form of life to arise on Earth. The reasons for such are clear:

- Prokaryotes are very simple. They do not have a nucleus or other organelles, and their DNA is not enclosed in a membrane. This makes them relatively easy to form, and it also means that they can survive in a wide variety of environments.
- Prokaryotes are very adaptable. They can survive in extreme environments, such as hot springs, hydrothermal vents, and the deep ocean. This suggests that they were able to adapt to the early Earth, which was a very different environment than the Earth we know today.
- Prokaryotes are the oldest form of life found on Earth. The oldest fossils of life on Earth are prokaryotes, and they are thought to be over 3.5 billion years old.

Therefore, it is likely that simple prokaryote-like lifeforms would be the first forms of life to arise in an ocean world and might actually be the dominant form of life throughout our universe. However, things become more fascinating once eukaryotes appear, leading to multicellular organisms.

Eukaryotes are divided into three kingdoms: plants, animals, and fungi. While plants and fungi are interesting, let us focus on animals which are characterized by locomotion and quick responses to external stimuli, as—let's be honest here—these are the types of alien lifeforms that most excite our imagination.

The animal kingdom is incredibly diverse and continuously studied by biologists to uncover new phyla (taxonomic rank) and species. There are 36 recognized phyla, of which but nine contain the vast majority of extant species:

- Porifera (sponges)
- Cnidaria (corals, jellyfish, sea anemones)
- Platyhelminthes (flatworms)
- Nematoda (roundworms)
- Annelida (segmented worms)
- Echinodermata (starfish, sea urchins, sea cucumbers)
- Chordates (animals with a nerve cord running along the back)
- Arthropods (segmented body animals)
- Mollusks (soft body animals)

The last three can be said to contain species with complex active bodies (CAB). These creatures possess the ability to swiftly navigate their surroundings, capturing and manipulating objects, and possess bodies that feature versatile appendages capable of moving in various directions, complemented by sensory organs like eyes that can monitor distant objects.

> Chordates are a large group that includes animals possessing a notochord at some stage in their life, which is a flexible, rod-like structure along the length of their body. Vertebrates, on the other hand, are a more specialized subgroup of chordates that have a backbone or vertebral column made of bones or cartilage, which replaces the notochord during development. In essence, all vertebrates are chordates, but not all chordates are vertebrates.
>
> Arthropodes are characterized by their segmented bodies, exoskeleton made of chitin, jointed appendages, and a high degree of mobility. They make up the largest animal phylum and include creatures like insects, arachnids (spiders, scorpions, ticks, and mites), crustaceans (crabs, lobsters, shrimp, and barnacles), and myriapods (centipedes and millipedes).
>
> Mollusks are a diverse and large phylum of invertebrate animals that encompasses a wide range of species, including snails, clams, mussels, octopuses, and squids. They exhibit a variety of body shapes, sizes, and habitats, making them one of the most diverse animal groups.

When studying the phyla that contain CAB, one starts to recognise commonalities. The most recognisable of these is of course a body plan which is a key factor in an animal's ability to survive and reproduce. The number of body plans found in nature is not fixed and can be difficult to precisely quantify due to the vast diversity of life on Earth. However, scientists have identified a range of fundamental body plans that organisms across different species share. These body plans represent the basic structural and organizational patterns that underlie various forms of life.

While there isn't a definitive number, some estimates suggest around 30–40 major body plans that encompass most multicellular organisms. We will review here the most common body plan variations to provide an educated guess on hypothetical lifeforms that might arise in subsurface oceans.

For a start, CAB animal species can be grouped by the symmetry of their bodies for which there are three categories:

> Asymmetry: Animals with asymmetry do not have any clear body symmetry. This type of symmetry is rare, and it is only found in a few groups of animals, such as jellyfish, sponges and placozoans.
>
> Radial symmetry: Animals with radial symmetry have a body that is arranged in a circular pattern around a central axis. This means that any dividing plane that passes through the center of the animal will divide it into two mirror-image halves. Examples of animals with radial symmetry include jellyfish, sea anemones, and starfish.
>
> Bilateral symmetry: Animals with bilateral symmetry have a body that is divided into two mirror-image halves by a single plane that passes through the head and tail. Bilateral animals have a front and back, a left and right, and a top and bottom. This type of symmetry is more common than radial symmetry, and it is found in animals that move around, such as insects, fish, and mammals. Most animals have bilateral symmetry.

Animals with radial symmetry are often sessile, meaning that they attach themselves to a surface and do not move around much. This type of symmetry is well-suited for animals that need to capture food from all directions, and it would also give them a more stable body plan in the turbulent water around hydrothermal vents for example. Animals with bilateral symmetry are more mobile, and they often have a head with sensory organs and a mouth. This type of symmetry allows animals to move around more easily and to capture food more efficiently. Animals with asymmetry are often very simple creatures, and they do not have the same level of mobility as animals with bilateral symmetry.

It is therefore very likely that extraterrestrial lifeforms will also follow a similar pattern according to their evolutionary history and its way of life. Mobile fish-like aliens that swim in an ocean will most likely have bilateral symmetry, while those that cluster around hydrothermal vents on the ocean floor will most likely have radial symmetry. Another commonality throughout animals grouped as CAB is the circulatory system. This is a complex network of vessels, a heart, and fluid (blood, coelomic fluid, or hemolymph) that collectively function to transport essential substances, such as oxygen, nutrients, hormones, and waste products, throughout the body. However, some aquatic animals that do not have a circulatory system, such as sponges and jellyfish which rely on diffusion instead to transport nutrients and oxygen to their cells. Diffusion is the process by which substances move from an area of high concentration to an area of low concentration. Animals without a circulatory system are typically small and simple, and do not need to move around very much, so they do not need a system to transport nutrients to their cells quickly. So, it is conceivable to envision uncomplicated jellyfish-like organisms leisurely drifting through an expansive subsurface ocean, gathering nutrients through diffusion as they go along. However, if an alien organism requires increased control over its surroundings, it is most probable that a circulatory system will be present. An additional common body arrangement involves the presence of a nervous system and sensory organs to inform it. Neurons and nervous systems most likely evolved on Earth among thin, motile, microbe-eating animals during the Ediacaran period (635–543 million years ago) and prior to the emergence of bilateral symmetry although this body plan introduced new opportunities with the evolution of the central nerve cord. Nervous systems are crucial for lifeforms that need to sense their environment and act upon it with appendages, which is why all animals possess one with the exception of sponges as they do not have any nerve cells or sensory cells. Thus, unless they are sponge-like, alien lifeforms will most likely possess a nervous system and sensory organs.The most common nervous systems found in the animal kingdom can be broadly classified into three categories depending on their structure and complexity:

Diffuse nervous systems: These systems are found in simple animals like jellyfish and sea anemones. They have a network of nerves that is spread throughout the body, and there is no central brain.

Ganglionic nervous systems: These systems are found in more complex animals like insects and spiders. They have a series of ganglia, or clusters of nerves, that are connected by nerves. The ganglia are responsible for coordinating the movements of the body.

Cranial nervous systems: These systems are found in vertebrates. They have a central brain that is made up of the brainstem, the cerebellum, and the cerebrum. The brainstem controls the basic functions of the body, such as breathing and heartbeat. The cerebellum controls balance and coordination. The cerebrum is responsible for higher-level functions, such as thinking, learning, and feeling emotions.

Again, according to the evolutionary history of an alien species and its way of life, it is most likely that one of these three types of nervous system will be present. After the establishment of a nervous system, organisms evolved the capacity to adjust to their surroundings, communicate, and engage in problem-solving. In other words, they evolved intelligence. Similar to 'life', intelligence is one of those words whose definition is a complex and controversial topic. There is no single definition of intelligence that is universally accepted by scientists and philosophers, yet most will agree that problem-solving, communication, and adaptability to dynamic environments are three prevalent indicators of intelligence.

Using these indicators, we rapidly recognize that intelligence arose independently within the phyla containing the largest, and most complex, lifeforms: chordates, arthropods, and mollusks, and that each of these approached the problem of intelligence in a unique way.

Apart from two small fish-like animals (lancelets and amphioxus), all chordates host a cranial nervous system which is characterized by a central organ where all cognitive tasks take place: the brain. This configuration has been widely successful as chordates span a wide diversity of animal intelligences from frogs to humans.

Arthropods also evolved their own form of central cognitive organ by clustering several ganglia allowing for complex problem-solving (some spiders use tools, such as twigs and leaves, to build their nests), however, the intelligence which arose from social insects, particularly those that live in organized colonies with specific roles and tasks, is distinct and demonstrates various forms of intelligence beyond individual behaviors. Referred to as insect social cognition, this form of intelligence enables certain arthropods like ants, wasps, bees, and termites to engage in complex problem-solving such as constructing intricate nests and traversing extensive distances. An extraterrestrial observer arriving on Earth 3 million years in the past might have found the intricate structures built by termites and ants more captivating than the developments of early hominins. Termites build mounds that are hundreds of times the height of an individual worker, and ants have well-maintained highways and elaborate nests. Early hominins, on the other hand, lived in small groups and foraged opportunistically from plants. They may have shown some group coordination in hunting, but they would have seemed like just another mammal to the alien visitor.

More intriguing though, is the way mollusks evolved intelligence. In addition to the clustering of ganglia forming a central cognitive organ like a brain, which allows scallops, for example, to communicate with each other using a system of clicks and whistles, certain types of mollusks such as octopuses also have neural clusters at the base of each arm. These 'mini-brains" are not as complex as the octopus's central brain, yet they are still capable of processing information, learning, and making decisions. In effect, these highly evolved animals have distributed their cognitive abilities into multiple locations throughout their bodies.

Even more remarkable though is the development of "higher intelligence" within the animal kingdom. This is generally referred to as advanced cognitive abilities and intellectual capacities that surpass basic or instinctual behaviors. As such, higher intelligence implies the capacity to understand intricate concepts, make connections, and exhibit creative thinking. Measuring higher intelligence is a complex task due to its multifaceted nature, however, it is well-established that many vertebrates, such as crows, dolphins, primates, dogs, and elephants, just to name a few, are considered to have higher intelligence.

With regards to arthropods, the question of whether insect social cognition is a form of higher intelligence in arthropods is still up for debate. Certain insects do seem to exhibit numerous elements of social intelligence akin to those observed in vertebrate societies, including individual recognition, learning object manipulation through observation, and even hints of cultural traditions (nest building in ants or dance patterns in honeybees). Insects also demonstrate states resembling emotions, potentially associated with social behaviors like coming to the aid of others in peril (red ants). Some scientists even speculate that specific instances of currently ingrained behaviors, such as constructing nests, might have originally stemmed from individual innovation followed by cultural diffusion, subsequently solidified by evolutionary processes into innate behavioral routines.

Surprisingly, mollusks have also evolved higher intelligence. Octopuses, which are considered to be the most intelligent mollusk after cuttlefish, have large brains, complex social structures, are masters of camouflage, capable of changing color, texture, and shape to blend into their surroundings, solve complex problems, are able to use tools, and also to learn from their experiences. Countless accounts of intriguing encounters between humans and octopuses in the wild and in captivity reveal that they have an inquisitive mind and rudimentary consciousness.

When viewed as such, it is interesting to note that all three animal phyla characterized by CAB have independently evolved their own distinct form of higher intelligence, or something very close to it in the case of insect social cognition. Thus, it is not too far-fetched to suggest that if we do discover alien lifeforms supporting complex active bodies within the depths of a subsurface ocean, some of these species might exhibit some form of higher intelligence.

Nevertheless, before we start imagining alien creatures living in complex social structures deep within Europa's global ocean, we must consider two crucial aspects which haven't been addressed yet.

The first one is the amount of energy available within a given system. Just like a thermal car runs on gas, life runs on energy; the more it has, the more it can do. As such, the availability of energy within an ecosystem on Earth is closely linked to biodiversity. Higher levels of energy input typically support greater biodiversity and allow for a larger variety of species. This energy availability enables the coexistence of various species with diverse traits, adaptations, and roles within the ecosystem. Conversely, ecosystems with limited energy resources may have reduced biodiversity due to constraints on the number of species that can be supported.

Overall, life on Earth has been lucky. Our planet is constantly bathed in solar energy, giving it ample fuel to try out the most fantastical of shapes and designs. Subsurface oceans are less fortunate. Scientists looking into this matter have proposed thermodynamic energetic calculations for subsurface oceans on Europa and Enceladus. These suggest that the energy available for metabolic redox reactions are comparable to Earth's low energy habitats such as the subseafloor oceanic crust, polar desert soils, and subglacial lakes. Thus, not the type of environment you would expect to see an energy-intensive complex active creature, let alone one an even more energy intensive with higher intelligence.

The second aspect which needs to be addressed when considering lifeforms in subsurface oceans is directly linked to the amount of energy available in a given system: the presence of oxygen. Because it is highly reactive (and potentially toxic for organisms that haven't adapted to it), oxygen is a great energy source; approximately a thousand more reactions can occur in an aerobic metabolism than in an anaerobic metabolism. In other words, oxygen supercharges the organisms that can use it. It is thus no surprise that eukaryotes and multicellular organisms arrived after the oxygenation of the seas, also known as the Great Oxidation Event, about 2.4 billion years ago, for their need for energy is far greater than simple unicellular organisms.

Therefore, when contemplating environments with low energy levels like subsurface oceans, it might be very likely that we encounter just a limited array of microbial organisms that are unable to support heightened complexity such as a cell with a nucleus.

However, it's important to keep in mind that these are speculations. The ocean worlds habitats are still poorly characterized and life has found many ways to surprise us in the past. So, who knows? Even if we discovered that the most complex organism in the vast subsurface ocean of Europa is a rudimentary creature no larger than the dot in this 'i', it would still be a biological revolution like none other.

In this section, we presented the rationale behind prioritizing the exploration of carbon-based life forms when searching for extraterrestrial life, and discussed the prerequisites, such as the need for liquid water, organic chemistry, an energy source like heat, minerals, and extended timeframes (though this remains a subject of ongoing debate). An intriguing prospect emerges where these four characteristics are believed to potentially exist within the context of subsurface oceans in certain icy moons and dwarf planets within our solar system.

Considering this, each ocean world candidate will be reviewed in detail and their habitability assessed whenever possible against these attributes. As such, we will delve into the specifics of the five icy moons where a subsurface ocean has been verified, the four planetary bodies (comprising two moons and two dwarf planets) that are highly probable to possess a subsurface ocean but are awaiting conclusive confirmation, and the numerous other celestial bodies that theoretically could harbor a subsurface ocean, even though limited evidence currently supports this notion. For each icy moon and dwarf planet, we will try to answer questions such as: What are the energy sources involved? What is the composition of the liquid mantle? How cold is the ocean? Are organic compounds present on the surface or in the liquid mantle, and if so which ones? Is liquid water directly in contact with the rocky material? How old is the body of liquid water? By scrutinizing these responses and evaluating additional attributes, we will formulate informed hypotheses about which ocean world candidate might be the prime contender for upcoming extraterrestrial life detection missions.

Now that we have gained a better understanding of the conditions necessary for life, we are poised to embark on our expedition across the solar system. The moment has arrived to secure our harnesses and signal mission control that we are prepared for take-off.

Part II
Confirmed Ocean Worlds

Where there's water on Earth, you find life as we know it. So if you find water somewhere else, it becomes a remarkable draw to look closer to see if life of any kind is there, even if it's bacterial, which would be extraordinary for the field of biology. (Neil deGrasse Tyson)

In this part, we review in detail the five moons that harbor vast subsurface oceans under their icy crust. Some are limited in their ability to support life while others boast the most promising environments for life to arise in our Solar System after Earth. All five are fascinating objects. Let us visit them one by one.

Chapter 4
Ganymede

Initial Approach

To reach Ganymede, we must travel to the Jovian system, home to the largest planet orbiting the Sun. On our approach, we encounter a bow-shock wave where the solar wind is deflected by Jupiter's magnetosphere, the largest and most dominant in our Solar System after the Sun's magnetosphere.

Jupiter requires respect.

As we get nearer to the planet, we are greeted by the Galilean moons. These moons hold together an impressive list of superlatives in our Solar System: the biggest moon (Ganymede), the most geologically active body (Io), the smoothest surface (Europa), the most densely cratered object (Callisto), and the only moon to have a magnetosphere (Ganymede).

The moon nearest to Jupiter, Io, is spewing significant amounts of matter out into open space—tidal heating at its most formidable—and although it is a geologically fascinating place to visit, there are no subsurface oceans there. Then comes Europa, Ganymede, and Callisto. As an ensemble, they form an excellent case study for subsurface oceans; Europa, the most promising in terms of habitability, contrasts well with Callisto, the least active of the three, while Ganymede, full of potential, lies in the middle, making it a good place to start our journey.

Ganymede Through the Ages

The first observation of Ganymede might have occurred more than 2000 years ago when, in 385 BC, a Chinese astronomer named Gan De noted in his records a bright companion to Jupiter. In the twentieth century, Xi Zezong, a renowned Chinese astronomer, asserted that this historical account provided evidence that the Chinese

© The Author(s), under exclusive license to Springer Nature Switzerland AG 2024   79
B. Henin, *Exploring the Ocean Worlds of Our Solar System*, Astronomers' Universe,
https://doi.org/10.1007/978-3-031-62953-2_4

were the earliest observers of the Jovian moon. However, Gan De refrained from making any definitive conclusions regarding the precise nature of his observation. Interestingly, he did note a reddish hue to the companion, which does not align with Ganymede's actual color, casting doubts on whether he indeed observed the moon. To Gan De's credit, he diligently conducted comprehensive astronomical observations throughout his lifetime. Furthermore, historical accounts have previously mentioned instances of naked-eye observations of a "star" positioned alongside Jupiter.

In fact, if it were not for its proximity to bright Jupiter, Ganymede—and the three other Galilean moons—would be visible in the night sky. Indeed, during Jupiter's opposition, when Earth passes between Jupiter and the Sun and is, therefore, closest to the Jovian system, Ganymede has a magnitude of 4.5, which puts it well within the range of the dimmest object the human eye can detect (magnitude 6). The key is to block the intense brightness of Jupiter, which induces spikes and flares in the human eye (a natural optical illusion), hiding any light reflected from the giant moons. Gan De used this exact method; his records show that he occluded Jupiter behind a tree limb.[1]

Regardless of who saw Ganymede first with the naked eye, it was Galileo Galilei who had the idea to point one of the first telescopes towards the Jovian system on that pivotal night of the January 7, 1610. In doing so, he was the first person to recognize that the bright companion was in fact a moon of Jupiter. There has been some dispute regarding this claim as Simon Marius, a German astronomer contemporary to Galileo, claimed to have discovered the moons a few months earlier. Unfortunately for him, as Marius kept no records of his observations, history sided with the Italian astronomer instead.

Galileo wanted to name the moons after his patron, Cosimo the Medici, the first of the famous Medici political dynasty but settled on using Roman numerical values with Jupiter I for Io, Jupiter II for Europa, Jupiter III for Ganymede, and Jupiter IV for Callisto. These became the common names of the moons up until the early twentieth century when they were replaced by the more familiar naming convention using mythological characters for which we must thank Marius (as well as Johannes Kepler, who convinced Marius that his original idea of naming Ganymede the 'Jupiter of Jupiter' was probably not the wisest idea). To this day though, you might still see the moons referred by their Roman numerical values.

Early on, Ganymede was thought to be one of the biggest, if not the biggest, of the known moons in our Solar System. We later learned that with a radius of 2634 km, it is the eighth biggest planetary object in our Solar System, closely

---

[1] Next time you find yourself far from light-polluted areas and Jupiter is high up in the sky, why not repeat the observation that Gan De did 2000 years ago and hide the planet behind a thin object (e.g., a branch) to see if you can detect the faint light of our Solar System's biggest moon. Make sure you check the real position of Ganymede afterwards with reference material to confirm your observations, as Jupiter could also be passing close to a faint star that might be confused with the moon.

followed by Saturn's moon Titan (2575 km), the planet Mercury (2439 km), and Callisto (2410 km).

Despite its massive size, very little was known of Ganymede before the advent of the Space Age and the flybys of the adventurous space probes in the 1970s and 1980s. Until then, astronomers were constrained by the vast distances that separated Earth from Ganymede and the technical limitations of their time as even through the world's most powerful telescopes, the giant moon was still just a bright speck of light. Despite this, astronomers designed ground-based observational tools that allowed them to better characterize Ganymede and other distant planetary objects; these tools included spectroscopy, photometry, radiometry, polarimetry, and radar. Let us review these one by one.

Spectroscopy takes the light emitted or reflected by a celestial body and splits it using a prism or similar optical device, thus providing the entire electromagnetic profile of the light, also referred to as the spectra. Within this profile are discernible patterns, manifested as absorption and emission bands, which can subsequently be matched against analogous patterns generated within laboratory settings. These matched patterns can then be deciphered to reveal insights from the object's surface and atmosphere.

In the 1950s and 1960s, spectroscopy had finally reached a sufficient level of maturity allowing astronomers to analyze the spectra of the Galilean moons. Early spectroscopic observations detected patterns of water-ice on Ganymede and Europa, correctly inferring that water was the main constituent of these moons' crust. It was also concluded back then that the moon's high albedo (the amount of sunlight reflected by a surface or atmosphere) was most likely caused by 'coherent backscattering in fractured ice'; in other words, the moons were most likely covered with ice. Due to technological advancements over the years, astronomers have been able to employ Earth-based spectroscopy to acquire a deeper understanding of Ganymede's surface composition, a practice that continues to this day. In fact, a recent study using this technique identified molecular oxygen and hydrated silicates composed of iron on the moon's surface.

Astronomers also utilize photometry as a valuable tool in their studies. It measures the albedo (the amount of sunlight reflected by a surface or atmosphere). For example, water-ice has a high albedo, as it will reflect much of the sunlight it receives back into space, while liquid water traps more light and has a lower albedo. By using theoretical models, astronomers can induce characteristics of a planetary object such as the existence or not of an atmosphere, the degree of topographic roughness (is the surface hilly or flat), or the nature of the surface material. Furthermore, in some circumstances, it is possible to observe a planetary body throughout its entire rotation showing albedo variations (and sometimes even color), suggesting different geological terrains across the whole surface. Photometry didn't show much variety on Ganymede but has revealed surface diversity on Europa, Io, and Dione.

Photometry is also used when a celestial body occults another body, such as when a moon transits in front of a star. The light captured at successive intervals during the occultation can reveal albedo variations from the surface, providing accurate measurements of the diameter or the detection of an atmosphere, as was the case for Titan. Unfortunately, the attempts to identify Ganymede's more tenuous atmosphere in the 1970s using this technique provided ambiguous results. It wasn't until 1995 when the Hubble Space Telescope began its observations of the giant moon that we detected an exceedingly faint atmosphere around Ganymede, commonly known as an exosphere.

In addition to photometry and spectroscopy, radiometry is the study of how planetary objects absorb or emit heat from the Sun. In essence, it measures the temperature (radiation emitted at thermal wavelengths). The critical factor here is the distance of the observed object from the Sun which allows astronomers to calculate the mean temperature of an object when the absorbed radiation is equal to the emitted radiation. Any observed temperature measurement that differs from this average can offer valuable insights into the characteristics of the object under study. Using this method, astronomers had already known that Titan and Io were emitting higher levels of heat than initially anticipated. The former was due to the greenhouse effect from its thick atmosphere, the latter from volcanic activity.

Even better, an ingenious method used in radiometry is the measurement of heat loss as a moon is being eclipsed by its main planet. This is called eclipse radiometry. A rapid heat loss from the surface during an eclipse is indicative of a porous surface that has difficulty trapping the heat. Such a porous surface is usually created by heavy meteor bombardments. When this technique was applied to the Galilean moons in the early 1970s, it was found that both Ganymede and Callisto lost heat rapidly, leading the scientists to presume that these two moons had heavily cratered surfaces, later confirmed by visiting space probes.

Polarimetry, another earth-based observational tool, measures the change in the polarization of sunlight being reflected by a planetary surface. The polarization will vary due to the shape, size, and other physical properties of atmospheric or surface particles. For example, our understanding throughout the decades of Titan's atmosphere has greatly benefited from this technique.

Finally, we have radar observations in which radio waves are targeted towards an object, and the echo is received back. The analysis of this signal can provide information on the diameter of the object as well as some insight into the surface composition. Studies conducted in the 1970s using the Arecibo Observatory in Puerto Rico accurately measured Ganymede's diameter and determined that its surface presented highly diffuse scattering, implying a rough and uneven surface. Such Earth-based observations did provide some insights on what to expect from Ganymede, yet with no surface images available; astronomers could only imagine what an ice-covered moon would resemble.

There was therefore great excitement when, in 1972, NASA launched Pioneer 10, the first mission to the outer planets, with the aim to explore the Jovian system and do a close flyby of its biggest moons. Being true to its name, the Pioneer 10 spacecraft had a number of firsts that have been hard to beat ever since: it was the first vehicle placed on a trajectory to escape the Solar System and venture into interstellar space; the first spacecraft to fly beyond Mars; the first to fly through the

Asteroid Belt; the first to fly past Jupiter, the first of the outer planets and the first to use an all-nuclear power engine to provide energy to its electrical systems.

And so, in December 1973, Pioneer 10 proceeded to the Jovian system, made its closest flyby of Ganymede at 443,000 km, and managed to take two fuzzy pictures of the moon. The quality wasn't optimal, due to the limitations of the spacecraft's rudimentary optical instruments,[2] and very little could be deduced with certainty, but light and dark surface patterns could be implied. We had to wait until 1979 when the more capable Voyagers sent back pictures revealing a genuinely intriguing world as can be seen in Fig. 1.4 in the first chapter. Upon reviewing these latest images, scientists concluded that the dark areas represented ancient surfaces covered with numerous craters while the lighter ones showed surprising signs of younger (but still old) geological activity with their grooves and ridges. It seemed that the newly discovered tidal heating process had visibly altered the surface of Ganymede but to a much lesser degree than what was observed on Io and Europa.

Between 1996 and 2000, higher-resolution images were provided by the Galileo spacecraft in orbit around the Jovian system, which made six close flybys of the moon. During its closest flyby, the second, the probe passed just 264 km from the surface, returning the most detailed surface images we have of the moon at the time of writing. It is worth mentioning that in addition to imaging instruments, the spacecraft mentioned (Galileo, the Voyagers, and Pioneers) also had remote sensing instruments such as spectrometers and photometers, which provided further insight into the giant moon's surface composition. These will be detailed below. Since then, however, various interplanetary missions have passed through the Jovian system onto other destinations (Ulysses, Cassini-Huygens, New Horizons, and currently Juno), yet none have come as close to Ganymede as Galileo did in the late twentieth century.

Ganymede's Story

Ganymede, along with the other Jovian moons, formed 4.5 billion years ago from the disk of dust and gas left over after Jupiter's formation. Due to the disk's density decreasing the further out you go from its centre, the more the mass of Jupiter's moons decreases outwards as well. The moon Io, the closest to Jupiter, has a density of 3.528 g/cm^3, while Europa, the second closest, is 3.014 g/cm^3; Ganymede, the third closest, is 1.942 g/cm^3, and finally, Callisto is the furthest and least dense with 1.834 g/cm^3. As a comparison, Earth's density is 5.5 g/cm^3 which is typical of a rocky planet. Since the density of water is 1 g/cm^3, we can already conclude that Ganymede and Callisto must be composed of a significant amount of water to bring their densities to such low values. And indeed, models currently estimate that water

---

[2] For more details on Pioneer 10's imaging instruments, or any spacecraft's imaging instruments, see the book titled "Imaging Our Solar System" by the same author and published by Springer International.

forms 46 to 50% of Ganymede's total mass. That is a significant amount of water for a planetary body that is bigger than planet Mercury.

Note that another consequence of the disk's decreasing density is that moons formed away from the giant planet take much longer to clear their orbits and reach their final masses than moons closer to Jupiter. Thankfully for Ganymede, its distance from Jupiter was still close enough to allow it to form within ten thousand years, fast enough to retain a substantial amount of primordial heat trapped in its core. Being still close enough to Jupiter also meant that the moon accumulated additional accretion heat, as the strong gravitational attraction of the giant planet substantially increased the impact velocities of any debris coming from outside the Jovian system.

On the other hand, Callisto took much longer to reach its final mass, therefore losing much heat in the process as well as experiencing smaller velocity impacts which contained less energy. This will have an important effect on Callisto, as will be seen in the following chapter.

Returning to Ganymede, it is thought that despite the moon's substantial mass, the energy from the primordial heat as well as the decay of the radioactive elements present in its rocky constituents were sufficient for it to undergo differentiation, the separation of the different components into distinct layers.

Also, it is thought that Ganymede might also have acquired additional heat early in its formation due to tidal heating. As you might recall from the first chapter, Ganymede's current orbit has an eccentricity, although it is tiny at 0.0013 (compared to Io with 0.0041 or Europa with 0.009). This leads to minimal tidal heating and negligible energy. However, some scientists speculate that the orbital eccentricity was much greater in the moon's past, as the three-body resonance with Io and Europa was being shaped, generating significant tidal heating. Once Ganymede was firmly locked into the resonance we see today, tidal heating was subdued and had less of an impact. If that wasn't enough already, recent studies have also speculated that Ganymede could have benefited from additional amounts of energy due to high intensity cratering in the Late Heavy Bombardment, although this is still subject to debate.

Regardless, we now know that Ganymede is fully differentiated with a solid inner core made of iron, an outer core of liquid iron and iron sulfide, a silicate mantle (where radiogenic heating is still occurring), and a thick outer layer of water in liquid and ice phase estimated to be around 800 km thick and containing up to 39 times as much water as our home planet. The exact way ice and liquid water is divided within the mantle is still up for debate, as various models have been put forward depending on the assumed composition of the water and other elements within the core, of which a few are presented here later.

Ganymede's liquid metallic core was discovered in the late 1990s thanks to the detection of a magnetosphere by the Galileo spacecraft, much to the surprise of planetary scientists. Galileo was equipped with two instruments: a magnetometer which measures the direction and strength of magnetic fields and a plasma wave spectrometer which measures variations in electromagnetic waves in the Jovian environment. At the time the mission was being conceived, both these instruments were selected to investigate Jupiter's large magnetosphere only.

However, when the spacecraft approached Ganymede, the plasma wave spectrometer detected an increase in charged particles by a factor of more than a hundred and the magnetometer sensed a sudden change in the magnetic field (increasing by fivefold). Scientists concluded that the only explanation for these measurements was that Ganymede had a magnetosphere, a first for a moon. And thus, scientists concluded that deep within the heart of the moon was nestled a liquid iron core with high electrical conductivity as only such a feature could generate the magnetic fields that had been discovered by the spacecraft.

Another first is that since Ganymede's magnetosphere is completely embedded within Jupiter's magnetosphere, the first known case of a magnetosphere within a magnetosphere. Thanks to its magnetosphere and its distance from the giant planet, Ganymede receives 450 times less radiation from Jupiter than Io (36 Sv or sieverts) and 68 times less than on Europa (5.4 Sv).

Surface Features and Exosphere

As speculated in the 1970s and confirmed by flybys, Ganymede's surface crust is mainly composed of water-ice. Due to this, there are relatively few big features such as mountains or high crater rims, as the ice present on the moon isn't strong enough to hold the weight of extensive vertical features. Apart from ice though, many non-water materials have also been detected on the surface.

Of these, salts have been observed in the form of magnesium sulfate and sodium sulfate, which might be similar to those found in the salty subsurface ocean. Although the thickness of the icy mantle separating the ocean and the surface is considerable (many hundreds of kilometres as we will see below), some scientists have speculated that these surface salts could be the result of brine making its way to the surface by eruptions or through cracks. Such active features have not been detected on the surface though.

Other non-water materials detected on the surface are organic materials that were most likely deposited by comets and other space bodies. These materials have been altered by Jupiter's radiation. There is also hydrogen peroxide attributed to photochemical reactions on the ice (in addition to the hydrogen peroxide found in the exosphere), sulfur whose origin can be traced back to Io, solid carbon dioxide (dry ice), and clays (mineral structures formed by organic matter and water).

However, as we have seen in chapter one, the surface of Ganymede presents a mix of two distinct terrain types: the dark and bright regions. The dark regions which comprise about one-third of the moon's surface, are ancient and heavily cratered. These contain the clays and other tarry organic materials mixed with the surface ice. Although craters can be observed all over the surface, the darker regions contain a higher concentration, implying that they are the oldest parts of the moon's surface. In fact, Ganymede seems to have undergone a heavy bombardment phase similar to what our Moon experienced around 3.5 and 4 billion years ago, so we assume that the darker regions date around from that epoch.

In contrast, the brighter areas are thought to be roughly 2 billion years old. These regions are covered with intricate patterns of long narrow grooves which lie parallel to each other in sets that can be hundreds of meters deep and extend far out to hundreds of kilometers, many from north to south. The origin of these grooves is still not fully understood. It could be due to either cryovolcanism or the remnants of past tidal heating. In both cases, interior convection would have been the culprit as warm, icy currents deep within the moon's interior would have strained the lithosphere, flexing and cracking the hard icy surface.

In a way, the darker regions resemble Callisto's heavily cratered surface, while the brighter regions are similar to Europa's young surface—although the number of craters on Ganymede's bright surface is still significantly higher than Europa's. In addition to the bright and dark surface regions, polar caps composed of water-frost were detected by the Voyager spacecraft. For many years scientists were intrigued by these features, although it is now thought that Ganymede's magnetic field has a significant role to play. Researchers have suggested that by funnelling the highly charged particles into the polar regions, Ganymede's magnetosphere had indirectly modified surface ice into layers of ice crystals, which brightens these areas. Since Ganymede is the only moon in our Solar System to have a magnetosphere, no other icy moon supports similar polar caps of bright water-ice.

Ganymede also supports a very tenuous atmosphere, referred to as an exosphere, and is mainly composed of oxygen. It is not unique for a moon to possess an exosphere like this; ultraviolet radiation from the Sun commonly breaks down surface water ice into hydrogen and oxygen gases. The former element quickly escapes into space due to its light mass and gets lost, while the latter is retained by the moon and forms the main constituent of the exosphere. Another resulting compound of solar radiation is hydrogen peroxide, detected mainly on the northern side by the James Webb Space Telescope (JWST) in 2023. Astronomers have also been able to detect through spectroscopy various gases trapped within the porous icy surface such as ozone (O_3), oxygen (O_2), and some small traces of hydrogen, all remnants of the broken-down water molecules. In addition to an exosphere, Ganymede supports an ionosphere as well. This ionosphere was confirmed in June 2021 when the Juno spacecraft flew past the moon and conducted a radio occultation experiment.

The Subsurface Ocean and Its Habitability

The existence of Ganymede's subsurface ocean as a distinct liquid water mantle was only confirmed in 2015 when to the Hubble Space Telescope discovered auroras while observing the moon.[3] (Fig. 4.1).

---

[3] It is worth pointing out that some unreliable sources have claimed that Ganymede's subsurface ocean had already been confirmed by the Galileo spacecraft in the late 1990s, but that is false.

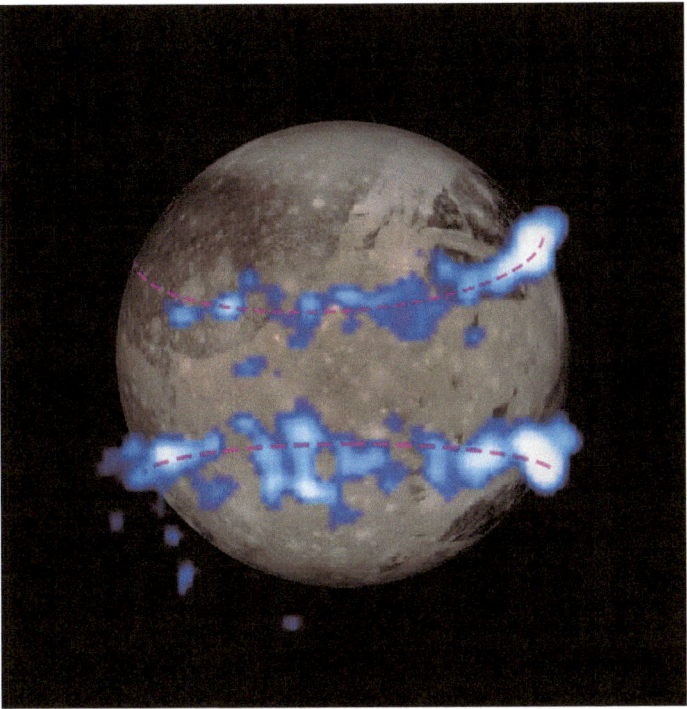

Fig. 4.1 Aurorae on Ganymede. The auroral belts can be seen in blue in this image by the Hubble Space Telescope. Their position indicates the existence of a salty subsurface ocean. (Image courtesy of NASA/ESA)

By observing the moon with its instrument sensitive to ultraviolet light, the Hubble telescope detected auroras similar to the ones we see on Earth. Aurorae on Ganymede are formed when the moon's magnetosphere forces high-speed subatomic particles from space to slam into the thin exosphere. The position of these aurorae will be determined by the interaction of Jupiter and Ganymede's magnetic fields. Thankfully we have a very good understanding of these. Given this, Ganymede's aurorae can be predicted using models of the moon's interior and magnetic fields.

Interestingly, the only model that fitted Hubble's aurora observations was not the one where the moon's subsurface water mantle was entirely made of solid ice, but was instead, in part composed of a deep salty ocean. This is because such an ocean will be able to create its very own magnetic field, albeit a very weak one, which will interact with Ganymede and Jupiter's magnetic field, which in turn influences where aurorae are formed.

It is currently estimated that, to explain the auroras we observe on the moon, Ganymede needs a globe-circling subsurface ocean that is at least 150 km deep and contains more liquid water than is found on Earth. Also, this subsurface ocean must contain salts, most likely magnesium sulfate and possibly sodium sulfate, as only a

GANYMEDE

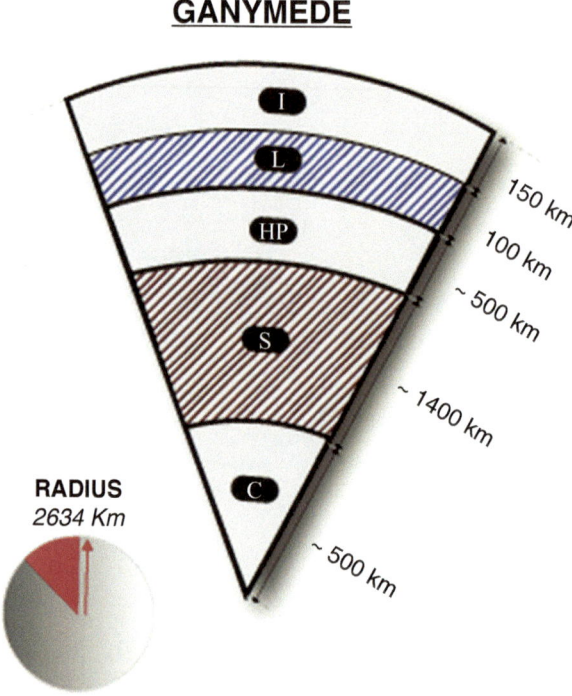

Fig. 4.2 Diagram showing the interior of Ganymede, according to the standard model where the subsurface ocean is sandwiched between two thick ice mantles. The thickness of the mantles is not well known, given the limited information we have on the moon. *I* Ice, *L* Liquid, *HP* High pressure ices, *S* Rocky, *C* core. Diagram is not to scale.

salty ocean can create a magnetic field strong enough to influence both Jupiter and Ganymede's magnetospheres. In addition, current models also suggest that the subsurface ocean rests under a very thick icy crust, most likely hundreds of kilometers thick. In this model, the subsurface ocean lies deep with Ganymede (Fig. 4.2).

The observations we have made so far in this chapter hold much promise for the habitability of Ganymede. Indeed, the moon has multiple sources of heat (radiogenic heating, primordial heating, and tidal heating) which have continuously kept the moon warm since its inception. It has also undergone differentiation with the result of hosting a mantle of liquid water in the form of a vast salty ocean circling the moon. Given what we know about Ganymede's formation, this ocean has existed for billions of years, more than enough time for life to start. Finally, organic compounds have been detected on the surface suggesting that such materials might also reside within the moon's interior.

However, there is a critical component that's absent: silicate matter. As you may recall from chapter three, microorganisms require minerals to facilitate redox reactions. In the case of Ganymede, this source of minerals will be the rocky mantle. Yet, is the subsurface ocean adjacent to it?

For decades now, the standard model of Ganymede's interior accepted by most planetary scientists proposes that the moon's subsurface ocean is sandwiched between two thick layers of ice: the icy shell forming the crust and the thick layer upon which the ocean rests. The bottom ice layer, in turn, sits on the silicate mantle. Unfortunately, in this model, rock and liquid water cannot interact.

The reason why another layer of ice is thought to exist under the subsurface ocean is due to the unique properties of water when exposed to extreme levels of pressure and temperatures. Let us explore these properties in more detail as this a recurrent theme throughout the models proposed for the other ocean worlds.

The state of matter of a chemical (liquid, solid, or gas) depends on its temperature and pressure as these directly affect how molecules arrange themselves. In a standard room where the temperatures and pressures encountered are adequate for our existence, water is in its liquid form. As we all know, at one atmosphere of pressure, water becomes solid ice if its temperature is lowered below 273 K (0 °C) and gaseous (steam/vapor) if its temperature is raised above 373 K (100 °C).

When water undergoes the transition into ice, it might appear as if it only exists in a single crystalline structure. However, subjected to high pressures and varying temperatures, water-ice can exist in numerous forms. As of today, we have identified 17 separate forms—and there may be more—known as phases, with each one being labeled by a Roman numeral in the order of their discovery. The existence of so many forms of crystalline water-ice is due to the unique properties of water, which no other molecule can match.

We need to understand the molecular structure of water (H_2O) in order to understand its chemical behavior. The oxygen atom possesses eight pairs of electrons, all of which repel each other. In a water molecule, two pairs of electrons bond with two hydrogen atoms, while the other two pairs of electrons are free. In effect, there are four 'items' sticking out from the oxygen atom, the two hydrogen atoms and the two pairs of electrons from the oxygen atom. The most stable arrangement for this configuration is the tetrahedron, with the oxygen atom at its center and the pairs of electrons furthest apart from each other.

This tetrahedron configuration forms a polar molecule because the oxygen end of the molecule is negatively charged, while the hydrogen end has a partial positive charge. In proximity to another water molecule, these charged ends form hydrogen bonds with each other; the negatively charged pole from one molecule bonds with the positively charged pole from another and vice versa. The strength of these bonds will depend on the physical properties of the environment in which the water molecules find themselves. They will either form loose bonds (liquid), strong bonds (ice), or little to no bonds (gas).

In its frozen state, water molecules are arranged with hydrogen bonds in a crystalline structure. Nevertheless, this compact arrangement causes electron pairs from neighboring molecules to experience repulsion. This creates the need for the molecules to find a stable crystalline form according to the physical properties of the environment they are in. And so, whenever the pressure or temperature conditions change, the water molecules—always seeking for stability—might shift from one crystalline form to another. To this day, we know of 17 different ways water molecules can structure themselves into ice.

By far, the most common of these phases found on our planet is called 'Ice Ih,' with the letter I standing for one (roman numeral) and h for holding for the hexagonal shape of the crystal. This is the one that falls from the sky as and brings endless joy to children as it is ideal for building a snowman or in snowball fights. Ice I (of which Ice Ih is part of) is the least dense of the water phases, which is why ice cubes float in your drink. As the pressures increase and the crystalline structures become more compact, the subsequent ice phases become denser and will sink. Another form of ice I present on Earth is called 'Ice Ic' which has a cubic crystalline shape; the letter c standing for a cube. This phase is found in high altitude clouds.

If pressures increase to 300 MPa and the temperature is at 198 K (−75 °C), ice Ih changes into a new crystalline form referred to as Ice II. This new phase of ice has a rhombohedral crystalline form. If Ice II gets warmed up to 250 K (−23 °C), it turns into Ice III, which has a tetragonal crystalline form. And as the physical conditions continue to change, fresh ice phases will emerge, each possessing distinct characteristics, until we reach Ice XVI.

All 17 ice phases have been created in laboratories on Earth by varying the pressure and temperature to which water is exposed to. We can even create forms of water-ice that exist in temperatures above the boiling point (such as Ice X). This is done by compressing the ice at extremely high pressures, compelling the molecules to stay stable regardless of the temperature.

In space where extremely low temperatures forces water-ice to form too quickly, ice doesn't have a crystalline form. Instead, this form of ice is referred to as amorphous ice as it lacks a crystal structure. This is by far the most common form of ice in space as it is found everywhere, from interstellar dust to the surfaces of comets, asteroids, planets, and moons.

Returning to Ganymede, the standard model of the moon's interior estimates the ice crust to be 150 km thick and the global subsurface ocean underneath it to be roughly a 100 km deep; as a comparison, the deepest point in our ocean is 10.9 km. The ice phase of this icy crust is thought to be Ice Ih. Models predict though that the immense pressures generated by the ice crust and the subsurface ocean, totalling a layer of water 250 km thick, are enough to coerce water at the base of the subsurface ocean to form an ice phase known as 'Ice VI' (a tetragonal crystal). Unfortunately for us, this results in the presence of an extremely thick mantle of hard ice lying between the subsurface ocean and Ganymede's silicate mantle, preventing any interactions between the liquid water and the rocks. Such a model limits any prospect for life as we know it.

Undeterred, scientists have published papers showing that magmatic events such as the movements of hot liquid rocks within the silicate mantle occurring at the interface between the rocks and the Ice VI mantle could generate pockets of water melts that slowly rise through the high-pressure ice (HP), carrying chemical nutrients and salts to the ocean above. Various characteristics such as the thickness of the ice mantle, the amount of heat exchanged, and the viscosity of the HP ice will affect the likelihood of this process; yet it does show that HP ices might not be a barrier to the transport of materials generated by water-rock interactions. It might be then that Ganymede's subsurface ocean receives 'drip-feeds' of volatiles and salts from below that would improve its prospect of being a habitable environment. Thus, a subsurface ocean sandwiched between two ice mantles might not be as isolated as we initially thought it to be.

Further hope came from a study in 2013 driven by Dr. Steve Vance from Caltech's Jet Propulsion Laboratory. Dr. Vance and his colleagues presented a new model, labeled 'club sandwich,' for the moon's interior that considered the effects of salt in the water mantles bringing the thermodynamics modeling closer to reality.

What the study suggested is unusual. Ganymede's interior might not be holding a single subsurface ocean layer but instead could be stacked by several layers of ocean separated by multiple layers of ice, hence the sandwich name. These ice layers would each be in different phases depending on the pressures applied to the ice (Fig. 4.3).

The ice crust formed by Ice I would be sitting on a thin layer of liquid water, which itself would be resting on another sheet of ice, this time in the phase Ice III under the form of snow; we will come to this shortly. This would in turn be resting on another thin layer of liquid water, which then would be resting on a layer of Ice V. Another thin layer of ocean would separate this layer of ice V to a layer of ice VI further below. At last, a final layer, this time of liquid water would be—hurrah!—resting on the rocky mantle. A warm salty subsurface ocean in contact with rocks. In other words; a potential habitable environment.

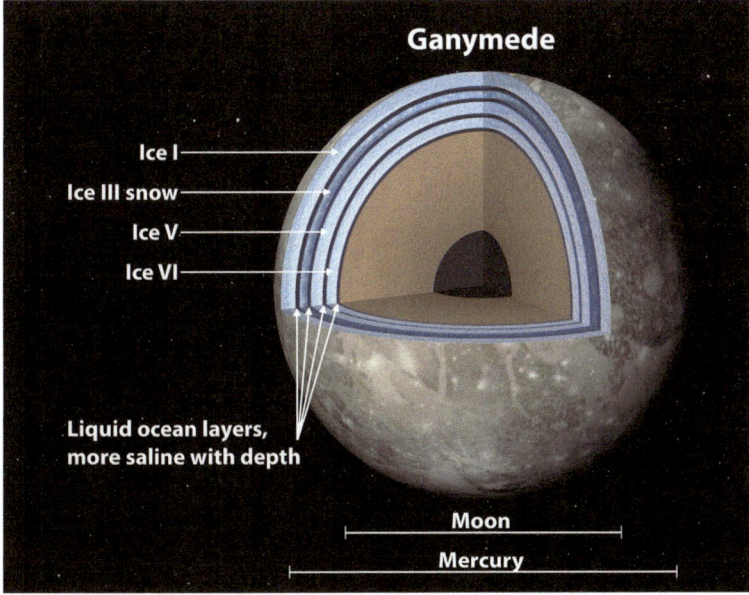

Fig. 4.3 Ganymede's interior as the "club sandwich" model, where multiple layers of liquid water are separated by ices. (Image courtesy of NASA/JPL and Caltech)

Prior to this new model, the studies were deliberately simplified as they didn't consider the presence of salt within the moon's interior. Salt is important though as it increases the density of liquids when exposed to extreme conditions such as those present inside Ganymede. Given how salt can modify the properties of ice and liquid water, it made sense to model Ganymede's interior with this in mind.

You can see the effect of salt on water for yourself. Take a glass of water and add table salt. Contrary to what common sense would assume, the level of the water will decrease. The explanation is simple: salt molecules will attract water molecules making the bottom of the glass denser and pulling the entirety of the liquid downwards.

Back to Ganymede, the 'lighter,' less salty, water mantle will sit on top while the saltiest and densest water mantle will sink to the bottom, making each liquid layer of ocean more saline with depth. As we go further down within the moon, the pressures, temperatures, and densities that can be found at specific depths provide the conditions for water to turn into ice on multiple occasions, which therefore creates the layering of ice and liquid mantles.

There is an interesting detail to this model: the emergence of a layer of Ice III in the form of 'snow' between the first and second layer of the liquid mantle. Indeed, ice can appear in cold churning waters such as in the liquid mantle. When salts precipitate out of the water and sink to the bottom, the ice there becomes lighter than the surrounding water and rises to the top. In other words, it will be snowing upside down.

In this sandwich model, the final layer of water adjacent to the silicate mantle is a liquid one: a subsurface ocean. Thus, we have water mixing with rocks. Also, we could reasonably assume that at such great depths, enough heat would be present to induce chemical reactions between the rocks and liquid water, although, given the extreme environment, determining what these reactions are is a difficult task.

This model is appealing, but is it conceivable, let alone stable? Can such a multi-layered structure last hundreds of millions of years? It seems difficult to answer. The complexities inherent to fluid dynamics make the equilibrium between the different layers challenging to demonstrate at present and may only occur under exceptional circumstances. Additional work is required. Future studies will no doubt provide new insights on the structure of Ganymede's water mantle as mathematical models are refined and new observational data acquired.

In a way, Ganymede has been lucky. It has continuously benefited from the energy generated by primordial and radiogenic heating and most likely also experienced strong tidal heating in its early phase, allowing for a significant melting of the water mantle. Do not be fooled if Ganymede doesn't portray itself as a very active moon compared to its youthful-looking neighbor Europa. The total amount of energy Ganymede holds might not be enough to keep the moon free from old craters, but it still offers enough energy to host a warm and extensive subsurface ocean.

A recent news about Ganymede's interior involved the Juno spacecraft currently orbiting Jupiter, and more specifically, JIRAM, the Jovian InfraRed Auroral Mapper spectrometer. As the spacecraft flew close to the giant moon on June 7, 2021, JIRAM detected mineral salts on the surface such as hydrated sodium chloride, ammonium chloride, sodium or ammonium carbonate, as well as organic compounds. The authors of the study concluded that these compounds most likely arrived at the surface from the extrusion of subsurface brines, and thus, reflect some of the possible chemistry within the subsurface ocean.

Moving forwards, a new spacecraft aimed at studying Ganymede in great detail is on its way: Juice (Jupiter Icy Moons Explorer) from the European Space Agency will orbit Ganymede in the early 2030s and give us new insights into this intriguing object. Further insights on Juice can be found in Chap. 12.

In the meantime, though, as we consider Ganymede's habitability, we can make basic assumptions. The standard model, where the moon's subsurface ocean is squeezed between two thick icy mantles, provides a liquid water environment, energy, and organic material as suggested in Juno's recent observations. As it stands though, life as we know it will not thrive in such an environment due to the lack of rocky materials. Nevertheless, this might change if it is regularly fed by minerals and salts from rising pocket melts. With the multi-layered sandwich model, the deepest and saltiest ocean layer adjacent to the rocky mantle seems to be a place where habitability might be conceived.

Regardless of which model is correct, we will most likely never know if any life is present in Ganymede's subsurface oceans. This is due to the extreme depths where these oceans are located. We should view this as a blessing. Ganymede is with Callisto one of the most likely locations for future human settlements in the outer Solar System once humanity develops the technological capability to leave Earth for new homes.

Indeed, Ganymede has much to offer for future colonists: a magnetosphere providing protection from Jupiter's wrath, water in seemingly unlimited amounts waiting to be used and converted into hydrogen, a small gravity that is not too strong as to make it energetically costly to leave the moon, but also not too weak as to make it challenging for its inhabitants, and finally easily accessible minerals (most likely metals) within Jupiter's Trojan asteroids (objects that share the same orbit as Jupiter). If humanity does step foot on Ganymede one day—and that is a big if—the subsurface ocean nestled deep within the giant moon will be completely sealed from the events taking place on its surface, leaving little chance for it to be disturbed by human activity.

We shall now leave Ganymede's tarred and fractured icy surface to visit its darker sibling, Callisto, as this moon also has much to tell us.

Chapter 5
Callisto

Callisto Through the Ages

As we approach the moon Callisto, a densely cratered landscape greets us. It's no surprise as the third largest moon in the Solar System is also the most heavily cratered. The biggest of these craters, Valhalla, is so big that it forms a planet-scale bullseye with concentric rings radiating for thousands of kilometers across the moon (Fig. 5.1).

Upon seeing Valhalla, one is immediately struck by the crater's immense size at 92 km wide. The moon's icy crust had no choice than to buckle and stretch to absorb the colossal amount of energy released by the impact. Callisto has tales to tell.

In terms of human observations, as opposed to Ganymede, there are no historical records of it ever being observed with the naked eye. Detected for the first time by the Italian astronomer Galileo Galilei in 1610, it was referred to as Jupiter IV for hundreds of years until it became known as Callisto, a name from Greek mythology chosen by the German astronomer Simon Marius.

As with Ganymede, observations of Callisto were limited during most of the previous centuries. It wasn't until the middle of the twentieth century with the arrival of more powerful observation instruments and techniques such as spectroscopy, photometry, radiometry, polarimetry, and radar that astronomers started to better characterize the moon. Earth-based observations showed Callisto to be an icy world due to its low density (1.834 g/cm^3), although the brightness observed from the moon's light was much lower than that of Ganymede or Europa, hinting that the icy surface had to be mixed with non-ice materials. Eclipse radiometry performed in the 1970s (see Chap. 4) detected a rapid heat loss from the moon's surface during an eclipse, implying that Callisto had a porous surface unable to retain heat; most likely due to deep layers of impact debris covering most of its surface.

© The Author(s), under exclusive license to Springer Nature Switzerland AG 2024   95
B. Henin, *Exploring the Ocean Worlds of Our Solar System*, Astronomers' Universe,
https://doi.org/10.1007/978-3-031-62953-2_5

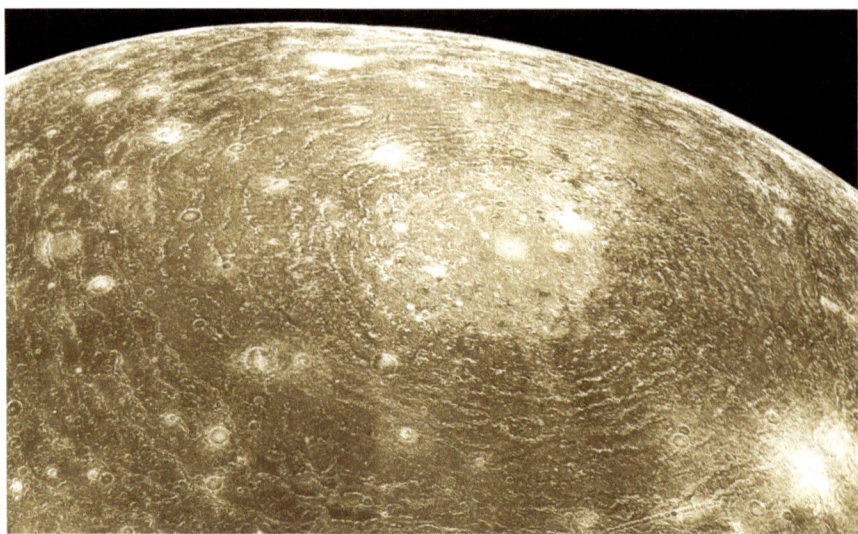

Fig. 5.1 Valhalla Crater, image taken by Voyager 1 in 1979. The ripples formed by the giant collision can be easily seen across the Callisto's icy surface. (Image courtesy of NASA)

When the Pioneer probes visited the Jovian system in the early 1970s, their instruments returned very little information about the moon. Instead, it was the two Voyager spacecraft that lifted the veil on this new world.

Images from the Voyagers with resolutions of 1 km per pixel revealed a surface covered with innumerable craters from top to bottom, with no trace of any past or present geological activity. With a crater density close to saturation, and any new crater erasing an older one, scientists quickly realized that Callisto was the most heavily cratered object in the Solar System. Apart from impact craters, relatively few surface features such as fractures, and escarpments could be found. With no mountains or valleys and the repetitive homogeneity of an intensely cratered surface, Callisto suffered from the comparison of its more visually appealing siblings. As such, it was known as the ugly duckling of the Galilean moons, or the dead moon (Fig. 5.2).

Nevertheless, the Voyagers returned only a limited amount of information during the flyby, and we had to wait 16 years for the next spacecraft, the Galileo probe, to arrive in the Jovian system. Galileo completed eight encounters with Callisto from 1996 to 2003, as opposed to 11 encounters with Europa, eight with Ganymede, and seven with Io.

With improved imaging instruments, Galileo revealed better surface details and a diversity of features. It also measured different properties of the moon such as gravity and magnetic data. Disregarded as dull after the Voyagers, Callisto was brought back into the light with Galileo's new insights as an ocean world and regained its glory as an object of major scientific interest. In 2000, another spacecraft, named Cassini-Huygens, flew by Jupiter on its way to Saturn and viewed

Fig. 5.2 Callisto in full view taken by the Galileo spacecraft in 2001. The moon's densely cratered surface is apparent in this image. (Image courtesy of NASA/JPL/DLR-Galileo)

Callisto using several of its instruments. New Horizons also whizzed by the Jovian system in 2007, but it was rather far away from Callisto so no high-resolution images were taken. Since then, there has been no significant visit to the moon, and much of the new data has been collected via Earth-based observations.

What have these missions taught us? For a start, Callisto took longer to form than the three other Galilean satellites, with estimates ranging from a hundred thousand years to ten million years; hardly a precise figure, yet still long enough to suggest that the moon struggled to capture some of the internal heat generated from its accretion. To make matters worse, since Callisto is located further away from Jupiter than the other Galilean satellites, its formation took place with lower-energy impacts. Furthermore, Callisto never experienced much tidal heating as it lies far from Jupiter's gravitational pull and wasn't locked into a strong resonance with its closer siblings, Ganymede or Europa. As a result, Callisto couldn't gather the energy required to undergo full differentiation.

With all this in mind, instead of being fully differentiated, Callisto's interior might be potentially full of slush of rocks and ices, with metals closer to the center, with some areas experiencing partial differentiation. We know this because scientists have devised a simple yet effective way of understanding what goes on under the surface of a planetary object. They measure its gravity.

Whenever the Galileo spacecraft flew next to Callisto (or any other planetary object), the Doppler shift of the spacecraft's radio signal would be precisely measured by Earth-based instruments, allowing NASA engineers to track with great precision the changes made in the spacecraft's velocity—the effects of the Callisto's gravitational pull—and with this they could extrapolate the mass variations across the moon. Since the mass is directly related to the internal composition of the moon and how it is structured, scientists could make educated guesses about its interior.

In addition, unlike its bigger sibling Ganymede, no magnetic field has been detected, implying that Callisto lacks a metallic liquid core big enough to generate a magnetosphere; another tell-tale sign that the rocks and metals in the interior never entirely separated.

Because of this, the moon's interior lacks the geological activity associated with a differentiated interior, such as the heat transfer from the convection within the different mantles, which explains why the surface's lithosphere has not changed since its inception. Whereas other icy moons have experienced significant resurfacing events, Callisto has kept large amounts of its crust intact for 4 billion years.

So, what have we found on the surface? Water-ice and rocks. No surprise here as the crust is a build-up of asteroid rocks and comet ices that have pummelled Callisto over billions of years. We also detected carbon dioxide ices in younger impact craters, sulfur which originated from Io's volcanoes, and various unidentified hydrated minerals that seem to have traces of iron as well as clays formed from magnesium. Tholins might have been detected as well (as they fit the absorption bands of 4.57 µm), which could have for origin the irradiation of methane and ammonia gases in the presence of water, carbon dioxide, and ethane ices. Did these compounds form on Callisto or were they deposited by falling space rocks? We don't know. It is a possibility that most of the molecules, organic or not, were formed outside of the Jovian system before being accumulating on the moon.

In a way, due to its lack of geological activity, the third biggest moon in our Solar System is a time capsule waiting to be explored. This is not entirely true though. The top surface has been exposed to the rigors of space and suffered numerous changes brought to it by external agents. Let us take a small detour to understand how the surface can be altered through time.

An Ever-Changing Surface

The first external agent to bring about physical changes to an exposed planetary surface are meteoroids, space rocks ranging in size from dust grains to small asteroids. When meteoroids strike a planetary body, they are called meteorites. A strike will form an impact crater that can alter the upper crust depending on the size and composition of the impactor. The more massive the object, the more material it will attract and the more energetic and consequential the impact will be. Thus, a giant moon such as Callisto will have brought on itself more impactors than smaller ones such as Dione or Mimas. A quick glance at Callisto's pulverized surface is enough to understand that the moon was subjected to innumerable impacts of all sizes and configurations. A striking example is Gipul Catena, a long series of impact craters forming a straight line, which was caused by an object that impacted Callisto in a similar way to the disintegrating Comet Shoemaker-Levy 9 striking Jupiter in 1994. It goes without saying that the presence of a thick atmosphere on a planet or moon can slow down and even neutralize small to mid-size ranging objects. The next agent responsible for modifying a planetary surface is radiation. In Callisto's case, Jupiter's powerful magnetosphere sends particles at high speeds across the Jovian system. Callisto's lack of magnetosphere, unlike Ganymede, exposes its surface to these high-energy particles, although, the radiation it receives is much smaller than what the three other Galilean moons are subjected to due to it orbiting Jupiter at a greater distance. Callisto receives 0.1 mSv per day. That's 800 times less than what Ganymede receives, 54,000 times less than Europa, and a whooping 360,000 times less than Io, the closest moon to Jupiter.

In a way, Callisto has the best of both worlds; the moon's location significantly reduces the effects of Jupiter's magnetosphere, while still allowing it to be shielded from most space radiations like solar and cosmic rays. Despite this, after hundreds of millions of years, countless high energy particles will have pummelled Callisto's surface in amounts sufficient to have altered its surface. This can be done in three ways: through chemical alterations, through the erosion of volatile components, and through the deposition of the particles (ions) themselves. Let us review each of these in more detail, starting with chemical alterations.

As we have seen with Ganymede in chapter four, the chemical reactions generated by high energy particles can break down surface water-ice into oxygen and hydrogen or modify it into H_2O_2 or H_3O. Additionally, oxygen and sulfur atoms can start combining and form SO or SO_2.

All these resulting compounds tend to darken Callisto's surface and, in some cases, even give it a reddish tint. As darker surfaces retain more heat, Callisto is the warmest of the icy Galilean moons with an average temperature of 134 K ($-139\,°C$) while Ganymede averages 110 K and Europa, the brightest, hovers between 50 K (poles) to 110 K (equator).

A powerful agent of change is ultraviolet (UV) radiation as it carries such a significant amount of energy (290–400 nm), that it causes changes to chemical structures of atmospheric and surface compounds. As an example, UV rays can create tholins in the presence of nitrogen (via ammonia) and methane. The presence of tholins is always an exciting discovery, as these organic polymers might become a food source for microorganisms given the right habitat. Unfortunately, since the Jovian system is located within nitrogen's frost line, the stable state of the nitrogen molecule will be gaseous, limiting the creation of tholins on the surfaces of Callisto, Ganymede, and Europa, although as we have seen, some small amounts might have been detected on these surfaces.

Luckily for us, our planet's ozone (O_3) layer shields us from most of these harmful UV radiations, but not all; this is why when it is recommended to apply sunscreen to your skin when exposed to the Sun. As such, multiple factors will determine how much UV radiation a planetary surface will be exposed to. Some of these include the presence of aerosols in the troposphere, the angle at which the sunlight reaches the surface as well as the elevation and reflectivity of the exposed surface, a thick atmosphere (including an abundant cloud cover), and the existence of an ozone layer in the stratosphere.[1]

Additionally, water in its liquid and ice forms offers one of the best protections against these harmful radiations, as water depth rapidly decreases, the exposure to UV rays decreases as well, especially when impurities and sediments are present. Callisto's tenuous atmosphere composed of carbon dioxide and oxygen resulting from the breakup of water molecules by radiation doesn't offer much protection, so the top surface gets its fair share of UV radiation, resulting in basic photochemistry such as the transformation of organic compounds and other non-ice molecules.

Finally, the last agent of surface change is the deposition of the dust and ice coming directly from space. Most of the major moons in our Solar System are tidally locked with their planets—referred to as 'primaries' in this case—which in effect, forms two sides to a moon: a leading side and a trailing side. The former, as its name suggests, is the side that is facing forwards as moon orbits around its primary, while the trailing side is the one facing backwards.[2]

In the same way as a car driving into a cloud of dust will gather more dust particles in the front windshield than in the back, a tidally locked moon will accumulate more space particles on its leading side than on its trailing side. A great visual example is Saturn's moon Iapetus which presents a much darker leading side in contrast to its brighter trailing side as it orbits Jupiter (resulting from debris located in its orbit which is being generated by Phoebe, another moon of Saturn). In the case of Callisto, the leading side is indeed darker—its trailing side has a higher albedo than its leading side—as it receives more micrometeorite bombardments and contains much more sulfur originating from Io's volcanoes.

Additionally, there is an agent of change that we have not yet mentioned here due to its negligible impact on Callisto: the galactic cosmic radiation or GCR. These are particles, mainly protons, that are accelerated to near the speed of light by stellar explosions located within our galaxy. GCRs are highly energetic and can significantly alter the molecular structure of a compound when struck. Thankfully for Callisto and the other Jovian moons though, Jupiter's powerful magnetosphere acts as a protective shield blocking the vast majority of GCRs.

Now that we have seen how the top surface layer of planetary objects can be modified by external agents, let us investigate endogenous agents. When, in the late 1990s, astronomers pointed Galileo's powerful cameras on Callisto's surface, they were expecting to see a multitude of impact craters of all sizes, from the tiniest to

---

[1] The formation of the ozone layer around 600 million years ago is one of the reasons why life on Earth was able to leave the protection of water and colonize the land.

[2] There will be no romantic Jupiter-rise for the future colonists on Callisto. You will be either on the near side where Jupiter is always visible or on the far side blanketed by a dark sky.

the biggest. However, few small craters with diameters less than 1 km were detected, leading scientists to come up with two explanations.

Firstly, smaller craters can be buried under a thick coat of ejecta; dark and powdery material generated from meteoritic impacts. In Callisto's case, countless impacts must have resulted in the production of vast amounts of ejecta. And indeed, some regions on the moon show elevated surface features such as crater rims poking out from thick ejecta layers.

There is another reason why small craters and surface features are less apparent on Callisto's surface: ice. When observing rims and central peaks from the largest impact craters, bright surfaces are often visible as the ices get exposed after an impact. Once surface temperatures reach 165 K (-108 °C), these exposed ices can sublimate, a process where the ices change directly into a gas, and they either leak into space or fall back, coating high altitude terrains, which in the process become brighter. High-resolution images from Galileo have revealed many imposing pillars of bright ice, often hundreds of meters high, which are being shaped by the steady sublimation, exposing fresher layers underneath. This process leads to a gradual loss of icy material that weakens surface features and leads to their collapse. The rims and central peaks of small craters contain less material and consequently undergo gradual erosion, whereas larger surface features will undergo alteration but still be visible. This process, known in geology as mass wasting and ground collapse, has been detected at multiple locations on Callisto. All these alterations might seem negligible when compared to other icy moons such as Ganymede, Europa, or even tiny Enceladus, which have all experienced significant resurfacing events, but they are nevertheless present on Callisto's surface.

Interestingly, there have been suggestions that features found in some regions on Callisto's surface could potentially result from tectonic activity. As such, five sites have been identified as displaying distinct linear features, such as narrow grooves resembling those found in Galileo Regio on Ganymede. It might then be that Callisto, in its youth, was subject to more tidal heating than what is currently expected, leading to surface alterations, the evidence of which was later erased through mass cratering and mass wasting. This is an intriguing idea, as it could be an indicator that the interior is more differentiated than we currently assume. Or it might be that these grooves are unrelated to tectonic mechanisms and formed through other means. At present, we just don't know.

Also, images taken by Galileo show flat darkish areas of limited scale that might be interpreted as cryovolcanic deposits, although there is currently no evidence for an endogenic process. Regardless, Callisto's simple geological history provides a good reference point for more complex worlds such as Ganymede.

Before exploring the moon's interior, we shall complete this picture of surface features by noting that Callisto has the fourth densest moon atmosphere within our Solar System—the other three are, in order of thickest, Saturn's moon Titan, Neptune's moon Triton, and Jupiter's moon Io. Being at fourth place, you might be tempted to imagine Callisto with a thick atmosphere upon which trailing clouds meander quietly, yet this is far from the truth. At 26 picobars, the atmosphere is very tenuous, being billions of times less dense than Earth's, and offers little to no protection from the outside elements.

However, this density still qualifies it to be classified as an atmosphere—as opposed to an exosphere, which is even less dense—as atmospheric molecules will bump into each other more frequently and create what we could call weather. Like Ganymede's exosphere, Callisto's atmosphere is mainly composed of oxygen that forms when water-ice molecules on the surface are split into hydrogen and oxygen atoms. Carbon dioxide is also present but at a very low concentration.

The Subsurface Ocean and Its Habitability

This seemingly uneventful moon is more interesting once we go underground. As surprising as it may sound for a partially differentiated object such as Callisto, we know that a global subsurface ocean resides deep under the icy crust.

This discovery was revealed in the late 1990s and early 2000s when the Galileo spacecraft made flybys of the Jovian moons and detected perturbations in Jupiter's magnetic field around Callisto (and Europa as well, which we shall see in the following chapter).

As you might recall from the previous chapter on Ganymede, the Jovian moons sit inside Jupiter's monstrous magnetosphere and—unless a moon has a magnetosphere of its own like Ganymede—these magnetic fields form a predictable pattern as the magnetosphere tilts up and down in relation to the moon's orbital plane. Because of these tilts, Callisto regularly experiences flips of the magnetic field as it orbits the planet. These flips can be predicted very well and were accurately measured by Galileo's magnetosphere. So far, so good.

Nevertheless, unpredictable variations of the magnetic fields were observed whenever the spacecraft was near Callisto, much to the surprise of the Galileo team. Something was interacting with Jupiter's magnetosphere, and after much speculation, the culprit was found: salt water in motion.

To understand this, we need to remind ourselves of a specific law of electromagnetism: a time-varying magnetic field will induce an electric field. This implies that an electrical current flows inside Callisto. This, in turn, creates a small magnetic field whose direction is approximately opposite to the primary magnetic field. The strength and response of this field can tell us a lot about the conductive medium located under the surface.

In Callisto's case, static icy mantles can't create the induced magnetic field responsible for the variability observed in Jupiter's magnetic fields. On the other hand, moving salty water can. Planetary scientists were excited about these results, especially since Europa, which was already known to have a global subsurface ocean, had produced similar types of electromagnetic variations. Could Callisto, the so-called dead moon, really have a salty subsurface ocean like Europa?

Feverishly, the scientific community worked on replicating the moon's internal structure through models that considered the gravity measurements taken by

Galileo, thermodynamic properties of the different states of water, high pressured ice as well as meteoritic material such as ordinary and carbonaceous chondrites, which are thought to be the moon's building blocks.

What came out of all this research is that models which most closely replicate the variability in Jupiter's magnetic fields host a global subsurface salty ocean. Thus, Callisto was shown to be an ocean world contrary to what might have been expected from a partially differentiated world.

More refined models now suggest that the ocean lies under 170 km of thick icy crust (in an Ice-I phase) and has a depth of at least 10 km but is most likely deeper. Since astronomers believe that Callisto lacks the heat input necessary to create and sustain this ocean, non-water material such as ammonia, salts, and other antifreeze components are thought to be present, as they make it easier to melt water at lower temperatures (although how these antifreeze materials arrived there is subject to debate). The models show that similarly to Ganymede, as we go down within the water mantle, pressures continue to build up and the liquid water changes into an HP ice mantle. Some models suggest that this mantle of HP ice is Ice-V and to be more than a 100 km thick. Underneath the Ice-V mantle possibly lies another layer composed of Ice-VI mixed with rocks, which itself is resting on an additional layer of Ice-VII also mixed with rocks. Further down and we encounter mixtures of rocks and metals. Other models suggest a much larger Ice-V mantle which directly rests on the rock and metal layers.

Given the uncertain nature of the moon's internal composition its rate of differentiation, it is important to remember that much remains unknown about Callisto's interior and that these models need to be taken with a pinch of salt (Fig. 5.3).

Regardless of which models we base our assumptions on, it is highly likely that under Callisto's subsurface ocean, there is a thick layer of ice mixed with some non-ice materials, sealing the ocean off from lower rocky/metal mantles, thus significantly reducing the ocean's habitability.

Furthermore, given the small amount of energy flowing into Callisto, there are far fewer chances that pockets of water melts rich in rocky material feed the subsurface ocean such as what has been suggested to occur on Ganymede. Therefore, Callisto's subsurface ocean seems to genuinely be sealed off between two very thick ice sheets with not enough heat to stir things up. As such, it is most likely to be an ancient ocean, unchanged and static since its formation billions of years ago.

Callisto is a fascinating world even when placed next to its more exciting siblings such as Ganymede or Europa. It has unique surface features and a surprising liquid interior. It is also the least studied of the Galilean moons, meaning that we still have much to uncover, and more surprises are likely to come. In truth, we understand far too little about Callisto and Ganymede to have any certainty about the nature of their subsurface oceans. It seems safe to assume for now that Callisto's habitability is very low and that its vast salty ocean is sterile. ESA's Juice mission (see Chap. 12), which will study Callisto in more detail in the 2030s, will be able to provide us with further insights on the subsurface ocean.

Ironically, one of the most interesting aspects of Callisto is not its lack of habitability for alien life but instead, its potential for supporting our own life, and more

Fig. 5.3 Diagram showing the interior of Callisto where the subsurface ocean is sandwiched between two thick ice mantles. The thickness of the mantles is not well known as well as the extent to which the interior is undifferentiated. *I* Ice, *L* Liquid, *HP* High pressure ices, *U* Undifferentiated. Diagram is not to scale

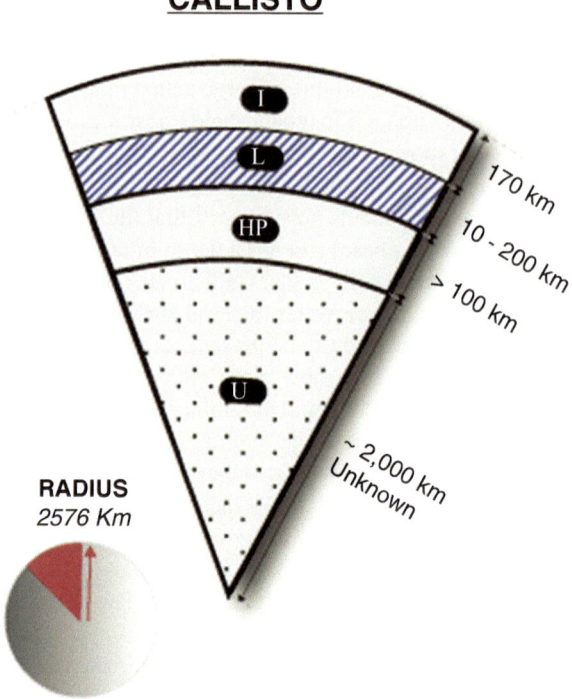

precisely, human colonists in the far future. Indeed, the giant moon hosts a range of conditions that make it attractive if we ever decide to set up surface habitats in the outer Solar System. In a paper published by NASA in 2003 under the title "Revolutionary Concepts for Human Outer Planet Exploration (HOPE)," seven authors selected Callisto as the best location for an outer Solar System colony due to its location (5 astronomical units from the Sun), its existing gravity (1/8th of Earth's gravity), its surface stability (due to the lack of geological activity), its abundance of water and other non-ice materials necessary for the production of fuel and life support systems, its relatively low exposure to Jupiter's extreme radiation environment, and its proximity to Europa allowing real-time teleoperation of robots to investigate Europa and Ganymede as well.

It is with this hopeful thought of humanity colonizing our Solar System that we depart from Callisto and visit the last ocean world of the Jovian system, and also the most promising of all the icy moons within our Solar System: Europa. There is much to explore there, so let's go!

Chapter 6
Europa

Europa Through the Ages

Arriving at Europa, the sixth biggest moon in our Solar System, we immediately get a sense for why, within a few decades, this moon became as alluring to astrobiologists as the planet Mars. From orbit, Europa is a perfect white sphere. In fact, it is one of the smoothest surfaces of any known solid object in the Solar System. We are far from the heavily cratered surface of Callisto, or the rugged terrains of Ganymede, as the tallest features on Europa are jagged 'blades' of ice which measure in the tens of meters, not hundreds. Standing on its surface, space travellers would observe a uniformly flat horizon everywhere they looked, like being on a giant snooker ball.

Intrepid travellers would nevertheless notice large areas covered with ochre patches and others laced with orangey-brown stripes. Despite these darkish features, Europa's surface is one of the brightest, with an albedo of 0.7. Ganymede has an albedo of 0.45 and Callisto 0.2. Only two other icy moons have higher albedos: Enceladus at 0.8 and Triton at 0.76. Since younger surfaces tend to reflect more light than older ones, we can already infer that Europa's surface is very young. A bright surface devoid of large structures can only imply one thing though: resurfacing events driven by recent geological activity. In other words, Europa is an active world!

As with the other Galilean moons detailed in the previous chapters, Europa was discovered in 1610 by Galileo Galilei, the Italian astronomer, and the confirmation of its status as an ocean world didn't fully occur until the Galileo, the spacecraft, characterized it in the late 1990s.

But let us go back to the start. Before space age exploration, various properties had already been inferred from Earth-based observations. In 1805, the French scholar Pierre-Simon Laplace managed the incredible feat of deducing its mass within 10% of its present value (4.7998×10^{22} kg). The diameter, far more difficult to establish, was provided in 1859 by Angelo Secchi, a brilliant Italian astronomer, with only a 6% error (present value is 3100 km).

B. Henin, *Exploring the Ocean Worlds of Our Solar System*, Astronomers' Universe, https://doi.org/10.1007/978-3-031-62953-2_6

With the moon's mass and diameter deduced, academics at the time could easily calculate its density, which is the mass divided by volume, and therefore make educated guesses about its composition. One of the first to try working out Europa's density was the prolific British amateur astronomer George Frederick Chambers. Unfortunately, basic errors in his calculations produced figures that were much lower than they should have been. Not only did he not realize his mistakes, but he also didn't provide any accompanying thoughts on the surprisingly low results.

Astronomers also tried to infer further properties by observing the variation of the moon's light (photometry, see Chap. 4), yet this method was so imprecise and fraught with errors that it gave way to inaccurate interpretations such as the belief that Europa was highly elliptical due to a fast-spinning rate. This theory stayed for half a century before being debunked.

We had to wait until 1908 when American astronomer, Edward Charles Pickering, made a serious attempt at working out the densities of the Galilean moons and interpreting their albedos to figure out their composition and structure. For Europa, he unfortunately also miscalculated the density—by a third lower—which made him suggest that it was composed of 'loose heaps of white sands' and 'dense cloud-laden atmospheres.'

After more accurate densities were estimated, an English astronomer and mathematician named Harold Jeffreys, was, in 1923, the first to suggest that due to their low densities, Ganymede and Callisto should be composed of icy materials as well as rocks, while Io and Europa should be mainly rocky since they had comparable densities to our Moon. Jeffreys' correct assumptions opened the door to a radical new idea, that rocky moons could be layered with ice, an idea that would become influential for decades.

New observations of Europa came about in 1927 when Joel Stebbins, an American astronomer, used photoelectric photometry, a technique he pioneered, to correctly deduce that Europa and the other Galilean moons were tidally locked with Jupiter, therefore always showing the same side towards the giant planet, just like our Moon.

In the early fifties it was assumed that while Io and Europa were probably rocky bodies, Ganymede was most likely a mixture of rock and ice composed of either water-ice or carbon dioxide, while Callisto was just a big chunk of ice. Ironically, few appeared to see a contradiction that while Europa and Io had similar densities to our Moon, their albedos were very different, and that lumping them into the same category might be misleading.

In the late 1950s and 1960s, new ideas introduced by Gerard Kuiper, the Dutch-American astronomer considered by many to have revolutionized planetary science, and other astronomers, combined with the arrival of new technologies, such as infrared spectral observations, led scientists to confirm that Europa and Ganymede did have water-ice on their surfaces.

In the 1970s, thermal radiation was measured for the first time on the Galilean moons. Unsurprisingly given its high albedo, Europa was found to be the coldest of the four at 120 K (−153 °C). This again confirmed that ices were present in great abundance on its surface, reflecting most of the Sun's heat. More accurate measurements, mainly during satellite eclipses (eclipse radiometry), led to the observation of temperature variations between the leading and trailing side, indicative of variance in surface features, as well as the suggestion that the surface was most likely made of low-conductive and porous material.

Around the same time, in 1971, an American astronomer, John S. Lewis, was the first to propose that planetary bodies could host an ocean of liquid water under an icy crust. The theoretical paper he produced was mainly based on the conditions that the water mantles would be rich in ammonia, and that radiogenic heating would be sufficient to provide enough energy to melt these mantles—tidal heating had not been conceived yet. Lewis even proposed that detecting such oceans might be possible with the presence of induced magnetic fields. Lewis was proved correct in the following decades (see Chap. 5).

All this was highly speculative though as our knowledge of our Solar System was very limited. At the time, Europa was still just a speck of light to the most powerful Earth-based telescopes.

Space Age Observations

In 1973, a new chapter in the exploration of our Solar System opened as NASA's Pioneer 10 spacecraft was the first human-made object to reach Jupiter and its satellite system. This was followed closely by its sibling, Pioneer 11, 7 months later.

Both spacecraft had for main objectives the measurements of fields and particles within the Jovian system, which was considered at the time to be more scientifically valuable than imaging planetary bodies. Nevertheless, they did carry imaging photopolarimeters that had three roles: analyze zodiacal dust, gain data on cloud particles in Jupiter's atmosphere, and preform photometry using red and blue filters. The instruments couldn't independently point at a target as they were fixed to the chassis, and instead, slowly scanned over an object by using the spin of the spacecraft. This method of data capturing would often introduce severe distortions that required heavy post-processing.

Furthermore, the photopolarimeters used by the Pioneers were limited by the technology of its time and had poor resolution capabilities. For this reason, the first images of the Galilean moons were crude and rudimentary. Also, during Pioneer 10's flyby of Jupiter at the end of 1973, Europa was far away in relation to its trajectory, and the probe managed to return just one image of the moon, which was fuzzy and difficult to interpret (see Fig. 6.1 below). Regardless, what had been for centuries a bright light in the sky was now a world to be studied.

The Pioneers also allowed astronomers to measure the mass of the Jovian moons with greater precision, forcing the scientists to revise the moons' densities and come to the realization that Io and Europa, with densities of 3.53 g/cm^3 and 2.99 g/cm^3, respectively, couldn't have the same composition and physical properties. This was the first time that the two inner moons were thought to be different.

Fig. 6.1 Europa viewed
by Pioneer 10, the first
picture taken of the icy
moon. (Image courtesy of
NASA)

In parallel to the Pioneer missions, theoretical modeling of the moons' formation continued apace due to the increased capabilities of computer simulations. In 1976, John Lewis and his student, Guy Consolmagno, published a seminal paper titled "Structural and Thermal Models of Icy Galilean Satellites," which provided a detailed hypothesis of the interiors of Europa, Ganymede, and Callisto. The paper illustrated the possibility of Europa possessing an icy crust measuring 70 km thick, with a liquid water ocean extending directly beneath it to a depth of a 100 km. Moreover, one excerpt from the paper made a remarkable prediction: "Europa would be more easily punctured by an impact: liquid water could then flow from the mantle onto the surface forming a flat, clean plain....". This was met with much skepticism from the scientific community, and papers were published in response to Lewis and Consolmagno's paper suggesting that subsurface oceans were unlikely.

Nevertheless, some scientists recognized early on the potential for Europa and other icy moons to harbor life if liquid water was present. A significant hurdle had to be overcome, though, as exemplified in a widely reported exchange in 1975 between Consolmagno and the famous astronomer Carl Sagan, where Sagan expressed his doubts of the idea of life in subsurface oceans, as life on Earth depended entirely on the light from our Sun, a source that was not available in distant oceans covered by kilometers of icy crust. He had a point. At the time, every life form known on Earth was linked one way or another to the Sun's energy output.

And then came 1977.

Like every good story, an unexpected twist occurred. In that year, a team of oceanographers took the scientific community by storm when they discovered on the East Pacific Ocean floor the very first chemosynthetic ecosystem: life forms living in total darkness within hydrothermal vents. Life, it seemed, could exist without the energy of the Sun after all. This breakthrough led biologists and astronomers alike to consider more seriously the potential of life within hypothetical subsurface oceans.

Thankfully, the scientific community didn't have to wait for too long, as fresh new data from the Voyagers 1 and 2 arrived in March and July of 1979. Due to orbital constraints, Europa was the least well photographed of the four moons during their visit, a reminder that the moon was considered less of a priority during the planning phase of the Voyager missions. Nevertheless, when Voyager 1 flew past Europa on March 5, it took the very first detailed image of the moon's surface, revealing a world of cracks and the notable red-orange lines crisscrossing the globe named lineae. It was assumed that these lines were due to plate tectonics. The absence of craters was very intriguing as well.

On July 8, Voyager 2 made the closest approach at the time as it passed at just 206,000 km from its surface and managed to return images with a resolution of 2 km per pixel. Although these maps were covering a fraction of the moon's surface, they made scientists realize that they were onto something unexpected as the bright surface was remarkably smooth, contrary to a world shaped by tectonic activity, and showed long linear markings that are similar to fractured sea ice (see Figs. 1.1 and 1.2 in Chap. 1). The lack of numerous craters or significant surface features implied that the surface was made of 'soft ice' incapable of holding a tall shape, while the linear cracks suggested a hardened crust becoming brittle under tectonic stress. Especially intriguing for the scientists at the time were strange features unique to Europa: the cycloidal ridges, which can be seen Fig. 6.2. Found near the moon's south pole, cycloidal ridges are symmetric double ridges forming sweeping arcs that run for hundreds of kilometers across the fractured surface. These bizarre features were not understood at the time, but most likely had to do with the way the surface was being deformed.

After reviewing the Voyager data, the position taken by the mission scientists was a conservative one. Europa was subjected to episodic heating, due to the newly discovered process of tidal heating (as mentioned in Chap. 1), which occasionally melted parts of the moon's thick icy crust.

The images returned by the Voyagers fascinated everyone though, firing the imagination of scientists and science fiction authors alike. Famously, at the time, the renowned science fiction author Arthur C. Clarke was writing the sequel to "2001: A Space Odyssey", and inspired by the Voyagers' recent discoveries, included an indigenous life form on Europa in "2010: Odyssey Two".

The Voyager's flybys of Europa also led to the involvement of NASA's planetary protection officer with the agency's next flagship mission, the Galileo spacecraft which would orbit Jupiter and visit its moons. The officer's role focuses entirely on preventing the contamination of terrestrial life forms with the habitable environments in our Solar System such as Mars and Europa, as well as ensuring that Earth's biosphere is protected in case life exists elsewhere. Even before Galileo arrived at Jupiter, NASA had already made the decision that the spacecraft should be destroyed.

Fig. 6.2 Cycloidal ridges
on Europa's surface
viewed by the Galileo
spacecraft in 1998. (Image
courtesy of NASA)

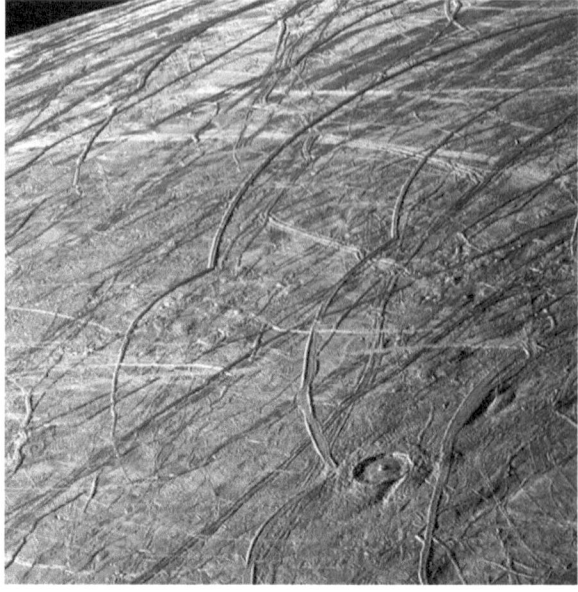

Fig. 6.2 Cycloidal ridges on Europa's surface viewed by the Galileo spacecraft in 1998. (Image courtesy of NASA)

The story of the development and launch of the Galileo spacecraft has become a classic tale on the dangers of politics influencing science. It is also the reason why we still know so little about the Jovian moons.

Galileo's Tale

Defined as flagship mission, the Galileo mission was initially labeled the Jupiter Orbital Probe (JOP). As such, JOP was already being conceived even before the Voyagers launched in the late 1970s to explore the Jovian system. Compared to the multi-planetary missions that were the Pioneers and the Voyagers, JOP had a deceptively simple objective: the detailed study of Jupiter and its moons. Alas, JOP proved to be everything but simple and would become a cautionary tale for future mission planners.

For a start, JOP required a multidisciplinary approach, as in many ways, Jupiter is like a miniature Solar System in itself with its collection of diverse moons, the presence of an intense magnetic field, the existence of swarms of dust and charged particles, and the giant planet itself lying at its center. And thus, the main challenge arose; the study of the giant gas planet, about which so little was known, demanded a very different approach to that of the Galilean moons. Due to this conflict of interest, planning trajectories and deciding what target should be prioritized over another proved critical in designing the mission.

Another major headache for mission planners was the fact that, due to this multidisciplinary approach, the mission was to be comprised of two spacecraft: the orbiter, which would weigh in at 2.5 tons, and the Jupiter atmospheric probe, which would weigh 339 kg. The scientific instruments planned for the mission would total 16 (in comparison, the Pioneers and Voyagers had 11), each collecting a fair amount of data despite limitations in the storage capacity of the spacecraft's central computer. Indeed, the large amount of data that would be collected by the orbiter and probe had already been identified as a severe bottleneck to the mission.

An additional challenge was the development of new technology designed explicitly for JOP, such as the first digital camera system (CCD) and the first imaging spectrometer ever to be flown into space. As such, JOP's scientific instruments represented at the time the most capable payload of experiments ever sent to another planet.

Additional complexity was added to this project when, in October 1977, it was agreed that the official launch date for JOP—which by then had been renamed the Galileo mission—was for January 1982, using the forthcoming, and still untested, Space Shuttle launch system. This proved to be an unfortunate decision.

The original plan for Galileo's launch was that once released from the Shuttle bay, the spacecraft would require a booster to take it out of low Earth orbit and place it on the required trajectory as the shuttles only reached an orbit of 320 km. The chosen booster was the newly developed Centaur-G, powerful enough to take the spacecraft on a direct course to Jupiter, ensuring a journey time of 2 years only. If all went well, Galileo was expected to arrive at Jupiter by 1984.

However, plagued by recurring and costly delays, the schedules of the Space Shuttle launches were continually slipping, and from Galileo's initial launch set for 1982, it was pushed back to 1984, then 1985, and finally to that fateful year of 1986, where Galileo was supposed to be launched by the Space Shuttle Atlantis.

Tragically, a few months before the spacecraft's planned launch, the Space Shuttle Challenger exploded during take-off, killing all seven astronauts onboard and grounding the Shuttle program for the following years. Galileo, whose mission was already delayed by 4 years, was forced into a storage facility next to the launch site in Florida and waited for a new launch date. Alas, that was not the end of its problems.

Indeed, the political fallout of the Challenger incident forced NASA to improve its safety regulations at all costs. The first victim of this new regime was the Centaur-G booster, which was deemed too risky as it involved carrying several tons of volatile liquid hydrogen and oxygen, which wasn't as tried and tested as solid fuel boosters. Some astronauts refused to take part in Shuttle missions if a Centaur-G booster would be present in the payload bay.

Faced with no other alternatives, the Galileo mission reluctantly ditched the Centaur-G booster for a more conventional solid fuel booster named Inertial Upper Stage (IUS), which unfortunately was much smaller and less powerful. IUS didn't provide enough velocity for Galileo to go on a straight trajectory to Jupiter; instead, a longer flight path had to be chosen to bring the spacecraft to Jupiter. This new flight path would use gravity assists, requiring two flybys of Earth and one of Venus. Contrary to Centaur-G's 2 years, using IUS would take Galileo almost 6 years to reach Jupiter. Worse was yet to come.

The new course had the spacecraft fly within the vicinity of Venus for a gravity assist which much closer to the Sun than what was initially planned, and since Galileo wasn't designed to withstand such high levels of solar radiation (a threefold increase), a total redesign of the spacecraft was required to protect its sensitive instruments.

Thus, Galileo had to be sent back to the other side of the American continent to the Jet Propulsion Laboratory in Pasadena, California, where it would stay there for 2 years as engineers added thermal shielding and made other modifications. Once the modifications were completed, Galileo headed back to Florida. Unfortunately, due to cost reasons, it was decided that the trip to Florida and back would be done on a flatbed truck.

Sadly, no one had realized that this cross-country journey on the American freeways would cause lubricant on some of the ribs of the spacecraft's primary antenna—the high gain antenna—to wear off completely. So, when Galileo finally successfully launched onboard the Space Shuttle Atlantis in 1989, 7 years after its intended flight, the mission was already compromised.

This came to everyone's attention when, on April 11, 1991, after almost 2 years in space and with the Venus flyby complete, the mission engineers instructed the high gain antenna to unfold its 18 ribs out from the central mast, which was designed to open up like an umbrella. Due to the missing lubricant, three or maybe four ribs refused to budge from the mast, and the whole antenna got stuck. It was half opened and tragically useless.

This was a major blow for the team and NASA. The high gain antenna was supposed to send data back home at a rapid rate to compensate for the computer's limited memory capacity. Without this capability, the mission was severely compromised. The billion-dollar flagship mission was in serious trouble, and this made the headline news.

Even after multiple efforts by the engineering team to address the issue, ranging from rotating the spacecraft at its maximum spin rate of 10.5 rpm to toggling the deployment motor on and off more than 13,000 times, the high gain antenna remained unresponsive, leading to the conclusion that the spacecraft could not be repaired. With no other choice, the spacecraft had to use the much smaller low-gain antenna, making the mission's data transmission rate abysmal. From 134 kilobits per second expected from the high gain antenna, the spacecraft was transmitting at 8 to16 bits per second. And while engineers managed, through software upgrades and data compression, to improve the transmission rate of the low gain antenna to one kilobit per second, it still represented a 99% drop in the data output initially planned for the mission. For many years to come, the Galileo team would be obliged to compromise on mission objectives to ensure maximum science return.

Irrespective of all the problems described above, another headache was to come in October 1995 while the spacecraft was on its way to Jupiter. The digital tape recorder, which stored the data before it was transmitted back to Earth, experienced a malfunction that damaged a good length of the tape located at the end of the reel. For precaution, the engineers sealed off a portion of the recording tape, constraining even further the data-collecting capabilities of the mission. This led to the decision to scrap planned observations of Io and Europa during the orbit insertion phase to ensure that the tape had enough space to store data collected by the atmospheric entry probe, which would plunge into the Jovian atmosphere.

All these problems left a bitter taste within the planetary science field. The irony of this story is that Galileo's hardships could have been easily avoided as the space-craft didn't have to fly on the Space Shuttle at all. During the conception phase, most of the Galileo team wanted the spacecraft to ride on Titan, an expendable rocket with a proven track record for sending payloads into space at a fraction of the cost of the space Shuttle and without any unnecessary risk placed on astronauts. Even better, the Titan rocket could also carry the Centaur-D booster, making the Titan-Centaur launch system far superior in every way to the Shuttle-IUS.

Alas, the politics of the U.S. space program had decided otherwise.

The ambitious space Shuttle program's development proved significantly more expensive than anticipated. To ensure its financial viability, NASA faced immense pressure to maximize Shuttle flights, transporting various payloads into low-Earth orbit regardless of their practicality. Requiring a crew of seven astronauts to put their lives at risk for a mission that a cheaper unmanned rocket could do better was highly questionable, but by then, the agency was burdened with the Shuttle program and forced to us it.

The Shuttle program did offer a new and promising way to bring payloads into low-Earth orbit and would allow for the construction of a permanent space station that would become the International Space Station. It just simply didn't make any sense to use the Shuttle for sending robotic payloads into deep space.

And so, a data-starved' spacecraft, old and fitted with 1970s technology, finally arrived in the Jovian system in December 1995, nearly 10 years later than initially envisaged. And yet, Galileo is still to this day, the spacecraft that made the closest approach to Europa and collected the highest-resolution data we have of the moon.

An Ocean World Revealed

The Galileo orbiter executed 12 close encounters with Europa during its three mis-sion phases. The first three close encounters occurred within its prime mission phase, from June 1996 to November 1997. Mission extensions were subsequently approved, allowing for an additional eight flybys during the Galileo-Europa mission phase (GEM) and a final close approach during the Galileo Millennium Mission (GMM), which ended in 2002. In addition to these close encounters, Galileo contin-ued to monitor Europa as it orbited Jupiter, albeit from far greater distances, and even though the images returned during these 'non-encounters' were not as detailed, they proved useful as they showed the moon in different angles and phases. At its closest approach, on December 16, 1997, the spacecraft passed above the surface of the moon at a hair-raising 201 km (lower even than the International Space Station's altitude) (Table 6.1).

Table 6.1 Galileo flybys of Europa during prime and extended missions GEM (Galielo Europa Mission) and GMM. (Galileo Millenium Mission)

Orbit name	Mission phase	Date	Altitude (km)
G1	Prime	27-Jun-96	1,56,000
G2	Prime	06-Sep-96	6,73,000
C3	Prime	04-Jan-96	41,000
E4	Prime	19-Dec-96	692
E6	Prime	20-Feb-97	586
G7	Prime	05-Apr-97	24,600
C9	Prime	25-Jun-97	12,00,000
C10	Prime	17-Sep-97	6,21,000
E11	Prime	06-Nov-97	2043
E12	GEM	16-Dec-97	201
E13	GEM	10-Feb-98	3562
E14	GEM	29-Mar-98	1644
E15	GEM	31-May-98	2515
E16	GEM	21-Jul-98	1834
E17	GEM	26-Sep-98	3582
E18	GEM	22-Nov-98	2271
E19	GEM	01-Feb-99	1439
I25	GEM	26-Nov-99	8860
E26	GMM	03-Jan-00	351
G28	GMM	20-May-00	5,93,321
I33	GMM	17-Jan-02	10,03,152

To capture pictureds of the moon, the imaging instrument on the Galileo orbiter—referred to as the solid-state imaging subsystem (SSI)—used a Cassegrain telescope coupled with a 176.5-mm aperture narrow-angle telescope that also included image sensors, focal plane shutters, electronics, and a filter wheel. It was developed for the needs of studying both the atmospheric motion on the gas giant as well as the Jovian moons. SSI's wavelength range was from the visible into the near-infrared, allowing it to identify different levels in Jupiter's atmosphere and geological formations on the moons.

The imaging campaigns required meticulous planning, as every image taken had to be sent to the tape recorder for temporary storage, which was later played back off the recorder and compressed by an onboard computer, until it could finally be sent back to Earth during cruise phases. Since the imaging instrument was a data hungry instrument, a large part of the storage capacity was being used whenever the spacecraft was taking images, limiting data acquisition for other instruments such as the spectrometer, the ultraviolet spectrometer, and the photopolarimeter-radiometer.

Another complexity arose from the two types of images required by the mission: regional views and close-ups. Regional views were provided by medium-resolution images that consisted of a few hundred meters per pixel, while high-resolution images at tens of meters per pixel, would allow the scientists to examine surface features up close. Ideally, both views would be taken from the same area as they complemented each other; the regional views giving context to the close-up images. This setup proved to be frustratingly difficult to implement throughout the mission as data limitations forced the imaging team to prioritize, with close views favored over regional views, making their interpretation more difficult.

To add further complications to an already challenging scenario, the position of the Sun relative to the moon would show different surface characteristics: the morphology of the terrain would be more visible in low-Sun angle views, while color images and photometry would be better served with high-Sun views. Each image taken by the spacecraft was the result of lengthy discussions and painful compromises within the Galileo team.

Nevertheless, the first high-resolution images of Europa's surface were taken during the prime mission phase (E6) on February 20, 1997, and acquired images of 21 m/pixel. These first-ever close-up images of Europa stunned scientists as they revealed a chaotic terrain full of ridges and displaced ice sheets that could be reconstructed together like a jigsaw puzzle (see Fig. 6.3).

On these images, Europa's surface was shown to be fractured everywhere as vast ice sheets jostled with giant titling blocks of ice resting on what seemed to be slush. Some even seemed to float—an impossibility since liquid water cannot exist in the vacuum of space. All this was intriguing.

In regions such as Conamara Chaos or Thrace Macula, disc-shaped areas were shown to contain 'floating' ice blocks that seemed to be stuck in a matrix of darkened material. These regions were labeled 'chaotic regions'. What was going on? At the time, the researchers weren't sure. Some suggested that Europa had a thin icy crust, so that liquid water from below would be very close to the surface, sometimes melting it, while others, still assumed that the moon had a very thick crust.

The images also revealed most compelling surface features found on Europa, the so-called lineae. These dark streaks covering the entire moon for thousands of

Fig. 6.3 One of Galileo's first high-resolution images of Europa's surface. Broken crustal plates seen here range up to 13 km across. (Image courtesy of NASA)

kilometers like a vast spider web are giant cracks where younger and brighter material seems to arise from the center, pushing the old darker material to the outer edges (much like oceanic ridges on Earth). This suggests a warming process that brings dirty, fresh slush/ice to the surface. Interestingly, a subduction process has been detected where plates of ice slide onto each other, in effect, analogous to tectonic plates on Earth, making Europa the only other planetary body in our Solar System where such geological activity has been detected.

Another essential characteristic of the lineae is their reddish-brown tint. Scientists weren't sure what to make of the odd coloring, as initial measurements on the nature of this non-ice material were unsuccessful. Still, to this day, various explanations have been put forward. One proposes that they are deposits of salts, coming from subsurface pockets of brine, which are altered by the intense radiation found on Europa's surface. Another explanation for the coloration of the lineae doesn't involve the subsurface ocean at all. As two segments of the icy crust buckle and rub against each other, a process referred to as shear heating occurs, where the two sheets touch, warming up the ice. This, in turn, sublimates the water in these specific areas, leaving enhanced concentrations of darker material that, most likely, would turn out to be sulfur from Io (more on this later in the chapter), therefore removing any need for a subsurface ocean contrary to the previous model. It might be that these two models of lineae formation occur at the same time. We currently don't know.

Galileo's first flybys also allowed astronomers to get a better idea of the moon's rocky interior; with an overall density of 3.01 g/cm^3, precisely measured by the orbiter. Thus, we know that the moon must be in large part composed of a thick, rocky mantle resting on a metallic core. At around a hundred to a hundred and 50 km thick, the water mantle consisting of the subsurface ocean and the icy crust is surprisingly thin in relation to Europa's mean radius at 1560 km and to the genuinely large water mantles under Callisto or Ganymede.

The Evidence of a Subsurface Ocean

Despite the data-bandwidth issue, the wealth of scientific data collected during Galileo's prime mission phase led to a two-year mission extension, the Galileo-Europa mission phase (GEM), whose primary focus was the study of Europa, and to a lesser extent, allow additional observations of Jupiter and Io. GEM consisted of three phases, each with a clear objective: the Europa phase was labeled "Ice", the Jupiter Water & Torus phase labeled "Water", and the Io phase labeled "Fire". No surprises, then, that GEM was also known internally as the "Ice, Water, and Fire" mission.

Consisting of eight close flybys of Europa, GEM's primary objective was to find further evidence for a subsurface ocean in the past and determine if it is still in existence today. This two-year extension was approved in 1997 by NASA and Congress—which ultimately holds the purse—yet it was done within the context of a cost-cutting period at NASA, the controversial 'Faster, Better, Cheaper' approach. As a result, GEM was only given 15 million dollars per year, less than expected, which required trimming spacecraft and ground operations to a bare minimum. Mission staff was cut by 80%, operational processes were streamlined and automated whenever possible, and data acquisition was severely restrained as only 2 days of data would be collected during close approaches of Jupiter or the targeted moons as opposed to a full 7 days during the prime mission.

Nevertheless, GEM proved to be a stunning success, as two lines of evidence were found for a subsurface ocean on Europa, one from numerous features imaged on the surface and another—more compelling—from the disturbance in Jupiter's magnetic field. Let us look at both lines of evidence in detail.

The first line of evidence is concentrated on Galileo's imaging data, where nine surface features were identified as consistent with a liquid water layer underneath the ice: impact morphologies, lenticulae, cryovolcanic features, pull-apart bands, chaos, ridges, surface frosts, topography, and global tectonics. It is important to note, though, that on their own, these geological features were not conclusive evidence, as they could also have been due to processes in warm, soft ice with only localized or partial melting. Only once scientists found evidence for an ocean independent from geological interpretation, did they become more confident in their understanding that the surface features seen on the images were also evidence of a subsurface ocean.

Impact morphologies, the first of the nine surface features indicative of a subsurface body of water, has to do with Europa's most prominent impact craters. On any planetary body, a crater's morphology can provide insights into the crust's physical properties, its composition, and potential depth. The study showed that the morphology of the biggest craters on Europa could only be explained if the icy crust is lying on a low-viscosity material, which in this case would be a layer of liquid water. Also, by analyzing 28 craters with a diameter larger than 4 km such as Tyre and Callanish, the two biggest craters on the moon, scientists have estimated an average crust thickness of 19 km—the crusts of Ganymede and Callisto thought to be ten times thicker.

The second set of features studied was the number of impact craters found on the surface. As a rule, the older a surface gets, the more craters it displays. Determining a precise age using crater counts for objects located in the outer Solar System can be tricky though,[1] and there is still much debate regarding the origin of some of the satellite systems orbiting some of the outer planets.

This hasn't stopped planetary scientists from working on models to estimate Europa's surface age, which they have put at around 40–90 million years. This is incredibly young by geological standards and suggests that the subsurface ocean must still be active at the present time.

---

[1] Contrary to the inner Solar System, where we were able to calibrate the major bombardment epochs with precise dates thanks to the Moon rocks returned by the Apollo missions.

The third study focused on large-scale fractures observed on the icy crust. More than a hundred of these faults were identified and analyzed, with a pattern emerging: the northern hemisphere is dominated by left-lateral offsets, while the southern hemisphere by right-lateral offsets. This gives us a clue to the compression forces upon which the icy crust is being subjected to through time. When comparing these patterns with computer simulations, the best match is that of the icy crust rotating at a different speed than the interior of the moon itself, an event called slipping in geology. This non-synchronous rotation of the icy crust can only be explained if the crust is lying over a fluid mantle, in other words, a subsurface ocean.

Furthermore, the cycloidal ridges first spotted by Voyager 2, were now understood to be formed from the tidal stresses generated by Jupiter, causing the subsurface ocean to ebb and flow similar to our tides on Earth. For each orbit, the water mantle experiences the rise and fall of tides by up to 30 m, inducing considerable stresses on the structure of the icy crust and forming arc-shaped cracks due to the orbital eccentricity.

Moreover, alongside this compilation of indirect geological indicators, scientists wanted to build a stronger case for a subsurface ocean. Prior to the last GEM flyby of Europa in February 1999, another mission extension was proposed. Named the Galileo Millennium Mission, or GMM, this extension had only one flyby planned for Europa (E26) while the rest of the focus this time was on the other moons such as Io, Callisto, Ganymede, and Amalthea, as well as the giant planet itself. GEM would end with the demise of Galileo in September 2003.

GEM was operated within the confines of another reduced budget, and a smaller team. But only one flyby of Europa was enough for scientists to finally confirm that, yes, the interference of Jupiter's magnetic field spotted during Galileo's first flyby in December 1996 (the E4 flyby) and subsequent encounters, was genuine.

The last flyby was thus performed in January 2002, and the detected interference became proof of an induced magnetic field generated by the moon in response to the periodic variation of Jupiter's magnetosphere. Europa's ionosphere was suggested as an explanation for these measurements; however, this explanation was deemed too weak to account for such robust currents.

What Galileo had detected could only be explained if a global layer of conductive material was located within the moon. As most conductive materials, such as graphite, were disregarded based on our understanding of the formation of Jupiter's moons and their internal composition, only a salty subsurface ocean could realistically match the observations. Subsequent models later revealed that this ocean must be lying 20 km beneath the icy crust and possesses a minimum depth of a 100 km (Fig. 6.4).

What level of salinity would be required in this ocean to produce the observed interference? We are not entirely sure as a range of solutions using different values for ocean depths, the degree of saltiness, and the ice crust thickness can match the measurements. Hopefully, new data from the ESA's Juice mission and NASA's Europa Clipper mission will allow us to have better insights.

The most promising characteristic of Europa's ocean, according to the models, is that due to its relatively low depth and the energy generated from tidal heating and

EUROPA

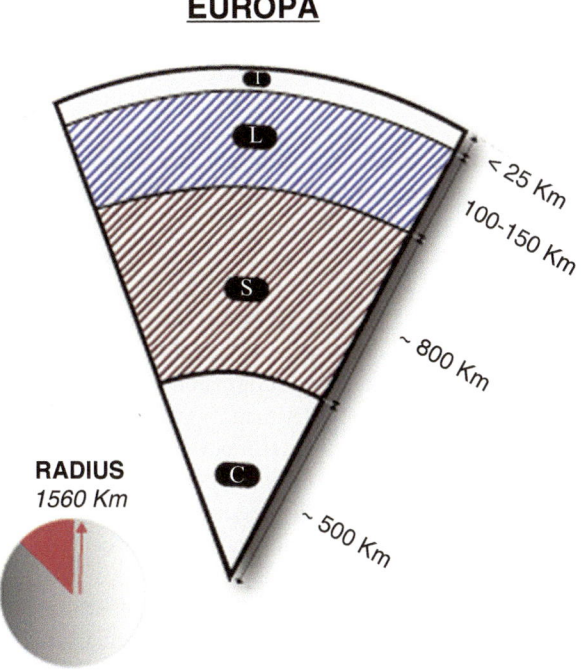

Fig. 6.4 Diagram showing the interior of Europa. The subsurface ocean rests on a silicate mantle where water and rock interactions occur. The thickness of the mantles is not well known. *I* Ice, *L* Liquid, *S* Silicate mantle, *C* Core. Diagram is not to scale.

radiogenic heating, no icy mantle separates the subsurface ocean with the silicate mantle below, allowing for constant interactions between liquid water and minerals.

Surface Features

The discovery of the subsurface ocean was only one of many scientific findings made by the orbiter. It also provided data on Europa's surface composition thanks to a near-infrared spectrometer (NIMS), a UV spectrometer (UVS), and a photopolarimeter (PPR). The results returned from these instruments found evidence that a thin layer of amorphous ice (<1 mm) was predominant on Europa's surface, revealing radiative disruption of the outermost layer of crystalline ice. This was mainly due to the high radiation environment within which the moon is bathed in. With time, radiation breaks down ice molecules in a process known as radiolysis, and this generates radiolytic products such as molecular oxygen (O_2) and hydrogen peroxide (H_2O_2). Impact gardening, which is the accumulation of rocky debris from micrometeorites that hit the top layer of the surface, can trap these molecules inside the regolith.

In addition to these two radiolytic products, Galileo's instruments also found carbon dioxide (CO_2), sodium (Na), potassium (K), sulfur dioxide (SO_2), and sulfur (S).

The origin of the carbon dioxide is not well defined. It could have been outgassed from the moon's interior or deposited by meteoritic material. On the other hand, the sulfuric material, comprised of sulfur dioxide and elemental sulfur, has most likely for origin the volcanoes of neighboring moon Io. We know this because these sulfurs are found on Europa's trailing side, implying that they mainly have an exogenous source. This might seem counterintuitive at first, given that it is the leading side that faces the forward motion during the moon's orbit around the planet. Therefore, one would expect the leading side to be covered with far more exogenous material contrary to the trailing side. Callisto and Ganymede are good examples of this as they both have a darker leading side compared to their trailing sides. So why is Europa's leading side generally brighter than its trailing side?

The answer to this puzzle lies in Jupiter's powerful magnetic field. While the immense planet rotates every 9.5 h, its magnetosphere synchronously rotates at the identical pace, transporting a plethora of particles along with it. As Europa takes 3.5 days to make a full orbit around the giant planet, it gets overtaken eight times by Jupiter's magnetic field. Similarly, this magnetic field sweeps past Io and strips away about one ton per second of volcanic gases and other materials, creating a large plasma torus around Jupiter and slamming sulfuric material onto the trailing side of Europa.

In addition to this, sulfur has also been found to correlate with certain geological features on the surface, suggesting that some of it could also be endogenic in nature. Future measurements will be required to confirm this.

Sodium and potassium were detected on Europa's atmosphere in 1996 thanks to Earth-based observations. Scientists believe these elements were lying on the surface as salts before being whipped up by Jupiter's radiation and ending up in the atmosphere. The exact origin of these elements is still not precisely known, but the main idea is that similar to sulfur, Io vents sodium-rich gases into space, which then slams into Europa. The sodium particles bore through less than a millimeter into the ice yet get trapped inside the fluffy regolith. After a while, though, the constant bombardment of particles raining on the surface erodes away the material covering the sodium, and these elements get released into the atmosphere.

Recent calculations, though, indicate that Io doesn't vent sodium in quantities large enough to explain the rate of detection in Europa's upper atmosphere. It therefore might be that some sodium is also generated by Europa itself, most likely through the radiolysis of sodium salts that must have been brought up to the surface as brines. Europa, therefore, might join the very small club of planetary objects that produce a net source of sodium, like the Moon, Mercury, and Io.

Unfortunately, the GEM and GMM extensions couldn't reveal the nature of the reddish material present on the lineaes and other surface areas. These are most likely composed of hydrated salts, although there have been suggestions that they could also be magnesium sulfate or even sulfuric acid.

Experiments conducted in laboratories on Earth demonstrated that when specific brines, such as water combined with magnesium and sodium salts, are subjected to conditions resembling those anticipated on Europa's surface, their spectra closely resemble those measured by the Galileo spacecraft.

Galileo's spectroscopic instruments were also unable to detect the nature of the substances that were coated with water. Referred to as hydrates, these elusive substances containing water are scattered across the surface. Earth-based telescopes have since carried on the investigation of the non-ice components on the surface with some success, as detailed further into this chapter.

When it comes to the mapping of the moon's surface, it was severely limited due to the low data bandwidth. Despite performing 12 close flybys between 1996 and 2000, Galileo couldn't map the entirety of the moon at high resolution. Instead, most of the surface underwent mapping with resolutions varying from one to 20 km per pixel, with certain regions enjoying a resolution of 200 m per pixel and some areas even reaching 10–20 m per pixel. Remarkably, one image was captured at 6 m per pixel making it the highest resolution image Galileo returned. A global map of Europa provided by the USGS astrogeology science center presents large unresolved swathes of land where the resolution is 20 km per pixel (mainly on the leading side). Compared to the jaw-dropping images of Pluto, Ceres, or Enceladus returned by modern-day spacecraft, it is unfortunate to note the absence of high-resolution mapping of one of the most captivating and significant moons in our Solar System.

Post-Galileo Discoveries

By the end of GMM, Galileo was in bad shape. Its two plutonium-powered thermo-electric generators (RTGs) were running out of fuel, and the lethal Jovian radiation environment had significantly limited the capabilities of the scientific instruments on board. Following a decision made years earlier, even before Galileo had reached Jupiter, the orbiter was ordered to plunge into Jupiter's thick atmosphere to prevent any risk from contaminating Europa given the fact that the spacecraft had not been sterilized,

And so, on September 23, 2003, Galileo burned in the upper atmosphere after spending almost fourteen years in space. No spacecraft has come as close to Europa since.

That hasn't stopped astronomers from persevering in their study of the icy moon. Following Galileo's demise, new observations have been made using Earth and space-based telescopes, and spacecraft passing through the Jovian system as part of a slingshot maneuver on their way to more distant targets within the outer Solar System. An example of this occurred in 2001, while Galileo was still in orbit around Jupiter. The brand-new Cassini spacecraft, on its way to Saturn, flew through the Jovian system to pick up speed. In doing so, it observed Europa from a fair distance with its ultraviolet imaging spectrograph (UVIS). UVIS splits ultraviolet light into its component wavelengths, allowing astronomers to identify atmospheric gases on planetary bodies. UVIS'results confirmed that Europa's atmosphere was much thinner than previously thought, a hundred times less than what models had predicted, reducing the likelihood that hypothetical plumes of water venting off from the surface might occurred on a regular basis.

In February 2007, another spacecraft, New Horizons, whizzed though space at neck-breaking speed and reached Jupiter in 13 months thanks to a powerful launch system combining an Atlas V rocket and a Centaur booster. On its way through the Jovian system, New Horizons observed in visible and infrared wavelengths the Galilean satellites, albeit at a far-off distance, thus providing images at resolutions of only 15 km per pixel. Nevertheless, these new observations showed that light scatters more homogeneously than expected, giving us some indication that Europa has a flatter surface than what had been previously estimated.

In September 2022, the Juno spacecraft came at 352 km of Europe's surface—nearly beating Galileo's closest flyby of 351 km in January 2000—and managed to make a swath north of the equator within 2 hours, returning images with the highest resolution at 1 km per pixel. Juno was never designed to be an imaging mission and only had an 11 mm focal length camera salvaged from the spares of the descent stage of the Mars Curiosity rover. It was the first close-up image of Europa's surface since Galileo and shows ridges and channels, impact craters and chaos terrains.

In addition to these distant flybys, planetary geologists continuously work on improving their mathematical models for how water-ice behaves in conditions like Europa. This led to the publication of a paper in 2011 titled "Active formation of chaos terrain over shallow subsurface water on Europa" which describes in detail how the chaotic terrains on the surface of the icy moon might be formed. The areas that have attracted the attention of scientists are the dark circular matrix of fragmented ice where 'icebergs' seem to be floating. New computer simulations have shown that these chaotic features can form with the rise of solid ice plumes, an upwelling of 'warm' ice behaving similarly to rocky plumes within Earth's mantle. Hot bubbles of warm ice forming at the base of the ice crust will steadily rise to the top, where its heat will dissipate. This process is known as convection and is an essential mechanism in distributing energy within the interiors of planetary bodies, rocky or icy. Picture a lava lamp with lumps of warmer material rising, and colder lumps sinking. The same process will occur in icy or rocky mantles, although it might take thousands or even a hundred thousand years for a lump of 'warm' ice to rise and fall.

As an ice plume rises through the icy crust, it can bring about substantial alterations to the material lying directly above it. When a warm ice reaches a certain height within the crust, it might have enough heat to melt the surrounding ice and create a subsurface lake near the surface.

Above the subsurface lake, the icy crust starts to weaken, allowing pressurized liquid water to flow upwards through the cracks and saturate the upper crust with heat. Once the heat from the 'warm ice' dissipates, the enclosed lake and chaotic terrain lying above it freezes up, causing the entire matrix to lift upwards like a dome (liquid water expands when turned to ice). The authors of this model have proposed the chaotic region of Thera Macula to be one of these active regions being formed and that we should observe noticeable changes between the Galileo encounter and a future flyby mission.

A final note regarding the chaotic terrains observed on Europa: they consistently appear soiled with non-ice material. This might imply that as the 'warm' ice rises through the crust, it brings contaminants from the subsurface ocean which then get mixed up with melted water nearer to the surface. If this scenario proves to be correct, we could have direct access to organic material from the subsurface ocean by exploring these chaotic terrains.

In 2013, new discoveries of Europa's surface composition were found using Earth-based telescopes. These discoveries demonstrate the rapid pace of technological innovations that have benefited astronomy since the Galileo spacecraft was built in the 1970s and 1980s. As such, ground-based telescopes have two big advantages over spacecraft whizzing across our solar system or even space-based telescopes. Firstly, they often lack restrictions on the size of instruments utilized. The necessity to miniaturize scientific instruments for space missions typically limits their capabilities. Secondly, they can be easily upgraded with the latest cutting-edge technology such as adaptive optics.

In 2013, spectroscopic observations made from Earth unexpectedly detected traces of magnesium sulfate ($MgSO_4$) on Europa's surface. This discovery holds significance as magnesium is a constituent element found in rocks. Back on Earth, magnesium is present in the rocky mantle—the eight most abundant element there—and is diluted in the oceans.

Hence, the identification of magnesium sulfate on Europa suggests that the subsurface ocean is directly interfacing with the rocky mantle. This interaction involves the dissolution of magnesium from the rocks and its transportation to the surface via the as-yet-unverified convection process within the icy crust. Once deposited on the surface, the magnesium interacts with the sulfur wherever it is present and forms magnesium sulfate.

One crucial element to support this theory is that magnesium sulfate has only been found on the trailing side, where sulfur is abundant. Because there is little to no magnesium sulfate on the leading side, we can assume that the compound wasn't created in Europa's interior. The discovery of magnesium sulfate on the trailing side also leads to another speculation. Due to its chemical properties, magnesium is never found unbound in nature. Instead, magnesium is always combined with another element, such as sulfur. Sulfur is contained in rocky material and gets dissolved in contact with water such as on Earth which has a sulfur cycle. However, the lack of magnesium sulfate on the leading side does imply that there just isn't enough sulfur present within the subsurface ocean to bind with the magnesium. Therefore, if sulfur didn't bind with magnesium on its way up to the surface, what other element might it have combined with?

In all likelihood, we should expect it to be chlorine. This element is also present in rocks, gets quickly dissolved, and can combine with magnesium to form the salt $MgCl_2$. Furthermore, as we saw earlier, previous observations have shown that sodium and potassium have also been detected on Europa. It is therefore likely that these two elements originated from within the moon. Bound to the chlorine as salts, these were transported to the surface as sodium chloride (NaCl) and potassium chloride (KCl).

This suggests that the subsurface ocean must harbor at least three types of salts: $MgCl_2$, NaCl, and KCl, and that chlorine should be present on the surface. Unfortunately, chlorine is not an easy chemical to detect using remote-sensing instruments. We will have to wait for future observations to finally confirm or not the existence of chlorine on the surface.

In the meantime, the discovery of magnesium, sodium, and potassium has opened up real possibilities that non-water material present within the subsurface ocean does indeed bubble up to the surface. This material could be sampled by a robotic lander. The simple idea has placed Europa as one of the best locations to search for extraterrestrial life. Another potential discovery could make it even easier for scientists to sample the moon's subsurface ocean.

Europa's Hypothetical Plumes

Motivated by the discovery, in 2005, of the existence of plumes at Enceladus' south pole (see Chap. 8), a team of scientists pointed the Hubble Space Telescope (HST) towards Europa and, to their surprise, spotted what they interpreted as an active plume located near the south pole. After additional observations, they proposed in a paper published in 2013 that the material venting out from the plume was water (inferred through studying the auroras on the moon) and calculated that the plume reached an altitude of 201 km before falling back down to the surface due to Europa's gravity (as opposed to Enceladus' plumes that are for the most part ejected directly into space).

The evidence of these jets was suggestive at best, and most within the planetary science community were skeptical. This is because the plumes being very faint, Hubble doesn't 'see' them, but instead, uses a transit technique to infer their presence; whenever Europa passes in front of Jupiter, the planet's ultraviolet light is precisely measured by Hubble. If something is dense enough to block the ultraviolet light, such as a plume of water venting off from the surface, it should be detected whenever the moon passes in front of the planet. This is what has been implied in the observations made in 2012.

Further detections were required, though, as the likelihood that the observation was a false positive was high due to instrument defects, and other factors that could introduce noise. It is with this in mind that researchers analyzed the raw data collected by the Cassini spacecraft as it flew by the Jovian system a decade earlier. Unfortunately, no water was detected in Europa's atmosphere or within the vicinity of the moon, excluding any plume activity at the time. If the plume phenomenon was genuinely occurring, then it must be intermittent. Luckily, more Hubble observations followed, and in 2014 and 2016 what seemed to be additional plumes were detected, this time from a region where the large Pwyll impact crater is located (Fig. 6.5).

In ten separate occurrences spanning 15 months, the HST observed Europa passing in front of Jupiter, and on three occasions, ultraviolet light was blocked in a way that strongly suggests the existence of plumes. Once again, the interpretation of this data seems to indicate that these plumes rise up to around 200 km before falling back down to the surface.

Two elements from these observations have reinforced the case for the plumes. The first is that the 2014 and 2016 detections were made at the same location: the region of Pyrll Crater, which reduces the likelihood that these measurements happened by chance such as an instrument fluke. The second is that this specific region on the moon's surface was observed by Galileo two decades earlier through thermal imagery and was identified as a 'hotspot': an area that is much hotter than its surroundings. The convergence of a hotspot coinciding with the obstruction of ultraviolet light strengthens the case for a plume. Either water from the subsurface ocean is venting outwards, or ice from the plumes fall back onto the surface, altering the structure of the top layer and making it better at retaining heat. Either way, these latest discoveries are intriguing and suggest the existence of plumes without confirming it.

Further mapping of Europa's surface has been made using the high precision Atacama Large Millimeter/Submillimeter Array (ALMA) based in Chile. When combined with complex thermal models, ALMA has created thermal maps of the moon and provided further insight on its thermal properties. This, however, has not shed further light onto the hypothetical plumes.

In 2019, a team at NASA's Goddard Space Flight Center confirmed the detection of water vapor above Europa's surface following 17 nights of observations between 2016 and 2017 with the Keck Observatory in Hawaii. This strongly indicates the

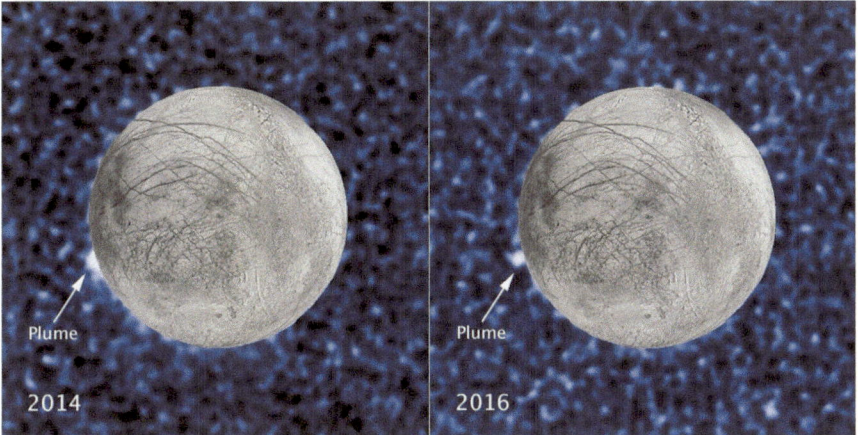

Fig. 6.5 Potential plumes on Europa detected by the Hubble Space Telescope in 2014 and 2016. (Image courtesy of NASA, ESA, W. Sparks (STScI), and the USGS Astrogeology Science Center)

potential activity of water plumes, although the observations only detected the presence of water vapor on one of the nights.

If future observations do confirm the existence of water plumes, the implications for the study and exploration of Europa will be substantial. Future spacecraft will be able either to fly through the plumes and analyze the ocean water at a safe altitude, or land on the surface and sample the material which will have rained back down. This eventuality, in addition to the direct sampling of the chaotic regions, as mentioned earlier, might remove the need to drill through Europa's icy crust to study its subsurface ocean.

Assessing Europa's Habitability

In terms of habitability, Europa seems to have the right set-up life: a vast salty subsurface ocean as ancient as the moon itself, around 4.5 billion years old, resting on the rocky seabed. In addition, the moon's eccentric orbit in resonance with Io and Ganymede generates inexhaustible amounts of tidal heating.

Of course, our knowledge of Europa is still very limited. If minerals are indeed being diluted into the ocean through the rock/water interface (more on this below), then the subsurface ocean should be composed of many more minerals than have been suggested so far. To illustrate such possibilities, an interesting study published in 2016 and led by Steve Vance from NASA's Jet Propulsion Lab, found that, chemical cycles within the moon could generate hydrogen and oxygen without the need for volcanic hydrothermal activity.

Indeed, hydrogen could be formed through a process called serpentinization, where salty water from the ocean seeps into the cracks within moon's crust and produce, via a chemical reaction, hydrogen and heat. Furthermore, models have shown that Europa's rocky mantle could have cracks as deep as 25 km, providing extensive surface areas for such reactions to take place around the entire moon.

As for oxygen, it would not be produced deep within the subsurface ocean, but instead, directly onto the surface. As we have seen earlier, water ice located onto the surface can be split by radiation into oxygen and hydrogen which escapes into space. This oxygen might be able to cycle back into the subsurface ocean through convection processes thought to be occurring within the icy crust. Further calculations have shown that oxygen could be produced at a rate ten times higher than hydrogen and might even reach levels that exceed Earth's oceans.

Would such an abundance of oxygen be suitable for life? Not necessarily. Oxygen is an oxidant, a substance that reacts quickly with other molecules, so structures vital for life would have a hard time assembling themselves in an oxygen-rich environment. Only after life has emerged can oxygen be gradually introduced into an environment, enabling the evolution of more complex life forms capable of tolerating it.

Similarly, given the ubiquitous nature of organic compounds in asteroids and comets, and the possible ability of Europa's thin crust to transport surface material to the subsurface ocean, it might be that such compounds could be present in the ocean as well.

Considering the aforementioned factors, some scientists posit the possibility of life emerging within the 4.5 billion years of Europa's subsurface ocean's existence. It might also be that geological processes have changed the properties of the ocean through time, probably making it more habitable at one period, and less at another.

A few more points are worth exploring. First, a big unknown for astronomers is where does the energy produced by tidal heating operate within Europa's interior? Could this energy be focused primarily on the surface where the icy crust meets the top of the ocean, resulting in the warming of both the top of the ocean and icy crust while keeping the seafloor relatively static? Or could the energy be instead directed towards the seafloor, where rocks would be heated up and chemically react with liquid water such as what occurs in hydrothermal vents.

Until we have a good idea as to where the heat from tidal heating is operating, we must work with the assumption that if life does exist on the moon, it could be located anywhere, from the cracks within the icy crust to the bottom of the ocean floor. It might even be that life thrives in the upper crust of the icy surface, protected from the harsh radiation by layers of water-ice—10 cm of depth would suffice to block most of the radiation—although the extremely low temperatures encountered throughout the crust would most likely prohibit life.[2]

Another point worth mentioning is that although there is much hope for the existence of hydrothermal vents operating on Europa's seafloor, there is no certainty that life could have started there. Indeed, even if these hot vents are one of the candidates for the origins of life on Earth, the extreme pressures encountered at the base of the Europan ocean floor, five times more than that at a mid-ocean ridge on Earth, would make the chemical reactions taking place within them unfavorably for life. For example, methanogenesis (the biological production of methane) is prominent within the hydrothermal vents on Earth as microorganisms have found a way to combine the carbon dioxide and hydrogen outgassed by the vents to form methane and water. This reaction—an equilibrium—is written like this:

$$CO_2\left(aq\right)+4\mathrm{H}_2\left(aq\right)\rightleftharpoons CH_4\left(aq\right)+2\mathrm{H}_2\mathrm{0}\left(\mathrm{l}\right)$$

At high temperatures such as those encountered within the vents, the equilibrium tends to be located to the left of this reaction, where most of the carbon is trapped as carbon dioxide. Life will shift this equilibrium to the right, where carbon forms methane instead, and in doing so, will capture some of the energy that is released.

This shift from left to right occurs naturally at lower temperatures. On the other hand, on Europa's ocean floor, where pressures are immense, this equilibrium would be strongly right-sided, making carbon dioxide less available for methanogens. In

---

[2] The coldest environment on Earth where active microbes have been found is $-20\ °C$ in the north Arctic. Cellular reproduction has been shown to stop after this.

such conditions, it would be difficult for life to sustain itself without viable an energy source.

It is nevertheless possible that other similar equilibrium reactions could be used instead, such as one using ferric iron instead of carbon. More research is required to simulate deep-sea vent environments lying on Europa's ocean floor, but it seems for now that biological methanogenesis is not the way to go on Europa.

Planetary scientists and astrobiologist got excited when the James Webb Space Telescope started a series of observations of the moon using its Near-Infrared Spectrograph (NIRSpec). This instrument mode provides spectra with enough resolution to allow astronomers to determine where specific chemicals are located on the surface. During these observations, carbon dioxide (CO_2) was detected in a large-scale region of chaos terrain known as Tara Region, as well as in Powys Regio. Given that these regions are most likely formed through the recent exchange of material from the subsurface ocean and the icy surface, it is probable that carbon originated from the ocean, revealing the potential for life forms based on carbon.

Even if life exists in the ocean, it might not be easy to find it. Highlighting this difficulty is the existence of a pelagic ocean back on Earth. A pelagic ocean refers to the open waterways that are far away from the shore or the seafloor and are characterized by a low density of life. Basically, in a pelagic ocean, life might be diluted to such an extent that discovering an organism would require a large amount of seawater to be sampled. Since Europa's ocean has a vast seafloor and is far deeper than any ocean on Earth, up to a hundred to a hundred and 50 km, there is a real reason to consider the limitations it could bring to our search for life.

Given the many uncertainties that remain regarding Europa's interior properties, a negative result in a life detection mission on Europa might not necessarily imply that there is no life in the subsurface ocean, just that we probably need more sensitive instruments.

Nevertheless, let's consider a scenario where fortune favors us, and life has indeed emerged somewhere within Europa's subterranean ocean. What life forms could be present there? Simple cells such as extremophiles on Earth (bacteria or archaea) thriving on chemosynthesis is the safest bet. More elaborate life forms such as soft-bodied jellyfish or hard-shelled shrimp would require a series of events such as the apparition of eukaryotic cells, multicellular organisms, oxygen-rich waters, as well as a constant source of energy capable of fuelling such creatures. That such events would have occurred both on Earth and within the interior of an icy moon seems highly improbable, but not impossible.

Although, those anticipating the presence of whale-like creatures in this ocean will be left disappointed. The evolution of life into complex organisms like large marine creatures necessitates intricate ecosystems, abundant food webs, and substantial energy inputs sustained over billions of years. Studies have suggested that, given what energy sources are known to be available within the moon's interior, complex life forms are highly unlikely.

Of course, nature can surprise us, especially when so little is known about an environment. Nevertheless, given our current understanding of biology and the known characteristics of the subsurface ocean, single-cell organisms are what scientists hope to find.

As such, the moon has become once again a prime target for space agencies. NASA's new flagship mission to the outer planets, the Europa Clipper, will be launched in late 2024 to investigate via remote sensing instruments the moon's habitability, while ESA's Juice mission will also visit Europa during two close flybys—see Chap. 12 to review these missions in further detail. Furthermore, the possibility of sending a lander to the moon's surface to do contact science and look for biosignatures is still being looked into, although not seriously proposed for now.

We must now turn our gaze away from Jupiter and its enthralling Galilean moons representing a remarkable assembly of celestial bodies within our solar system, and set our sights on another realm of equal intrigue: the Saturnian system.

Saturn, the second among the gas giants, boasts not only its spectacular ring system but also harbors a vast and varied collection of satellites. Among them, three exceptional moons stand out, Titan, Enceladus, and Dione, each holding promise for exploration in our quest to understand ocean worlds. The next two chapters will be devoted to the confirmed ocean worlds of Titan and Enceladus, thus completing the second part of this book, while Dione will be covered in the third part.

Let's go to Saturn!

Chapter 7
Titan

The Saturnian System

Surrounded by an abundance of moons and rings, Saturn is home to some of the most awe-inspiring vistas in our Solar System. Its satellite system is a rich one, hosting a varied collection of moons, sometimes referred to as the Cronian moons. (Krónos is the Greek name for Saturn.) At the time of writing, a total of 146 had been confirmed so far, with another two provisional. These include Titan, the second largest moon in our Solar System, and an assortment of mid-size to small icy moons each unique and intriguing, listed here from biggest to smallest: Rhea, Iapetus, Dione, Tethys, Enceladus, and Mimas. Saturn also has tiny irregular moons such as Hyperion, Phoebe, and Janus. Amalthea, Jupiter's fifth biggest moon is smaller than Janus.

Because Saturn's orbit lies much further out from the Sun than Jupiter, icy compounds are found in greater abundance. Nothing illustrates this better than the planet's majestic rings, which are made up almost entirely of tiny particles of water-ice. Also, many of the moons have significant amounts of water such as Tethys and Mimas which are thought to be made up almost entirely of water-ice. Antifreeze compounds are also present in the Saturnian system, making ocean world candidates more likely (Table 7.1).

The Saturnian system is structured as follows: closest to the gas planet lies the classical ring system labeled from A to F and the tiny moons Prometheus and Janus; these are then closely followed by a string of icy moons that increase in size as we move further out: Mimas, Enceladus, Tethys, Dione, and Rhea. Sitting relatively close to one another, these moons have at times come in and out of resonance with each other. A vast, diffuse ring, the E ring, is embedded within the orbits of these five moons, with Enceladus for its source (see Chap. 8).

As we move further out from Rhea, the furthest of these inner moons lying roughly at half a million kilometers from Saturn, we encounter a gap of 700,000 km

B. Henin, *Exploring the Ocean Worlds of Our Solar System*, Astronomers' Universe, https://doi.org/10.1007/978-3-031-62953-2_7

Table 7.1 The main characteristics of Saturn's seven biggest moons.

Moon	Mean distance from planet (1000 km)	Orbital period (days)	Mean radius (Km)	Mass (10^{20} kg)	Density (10^3 kg/ m³)	Eccentricity	Year of discovery	Name of discoverer
Mimas	186	0.94	199	0.38	1.15	0.0202	1789	W. Herschel
Enceladus	238	1.37	249	0.73	1.61	0.0047	1789	W. Herschel
Tethys	295	1.89	530	6.2	0.96	0.0001	1684	G. Cassini
Dione	377	2.74	560	10.5	1.47	0.0022	1684	G. Cassini
Rhea	527	4.52	764	23.1	1.23	0.0013	1672	G. Cassini
Titan	1222	15.95	2575	1346	1.88	0.0288	1655	C. Huygens
Iapetus	3561	79.3	718	15.9	1.09	0.0286	1671	G. Cassini

where no moons or rings can be found. We then reach Titan, orbiting at 1.2 million kilometers from Saturn. The giant moon has for close neighbour the tiny moon Hyperion, which is so small that it can't even form a spheroid shape. Then, mid-size Iapetus is located at three times the distance Titan is from Saturn, while the next moon after that is Phoebe, located even further at ten times the distance (notwithstanding the two moonlets Kiviuq and Ijiraq as well as the Phoebe ring).

Titan will be examined in this chapter, Enceladus in the next, Dione in Chap. 9, and Rhea in Chap. 11, while Mimas can be found in the appendices section of this book.

Titan's Discovery

The discovery of the second biggest moon in our Solar System stands as a timeless tale within the annals of astronomy. Discovered by Dutch astronomer Christiaan Huygens on March 25, 1655, Titan had actually been spotted earlier by the Polish astronomer Johannes Hevelius. The first Pole to be made a member of the Royal Society in London, Hevelius's second wife, Elizabeth Koopman, is one of the first professional female astronomers. Prior to Huygens' observations, Hevelius had studied Saturn for 14 years, from 1642 to 1656, and made countless drawings of strange sickle-like shapes around it. These were named 'ansae'as astronomers at the time didn't recognize them as rings. During his observations of these ansaes, the Polish astronomer would spot now and again a 'star' close to Saturn, but not knowing what to make of it, considered it to be just another star (star charts for telescope observations were rare at the time) and didn't mention it to his peers.[1]

Huygens, on the other hand, understood very quickly that this 'star' might be something else. An accomplished mathematician and burgeoning astronomer of his

---

[1] In a similar fashion to Helevius, the British astronomer Christopher Wren was also thought to have spotted Titan earlier but had failed to make the connection between the 'star' and a potential moon.

era, Huygens was eager to personally observe the enigmatic ansae of Saturn that were being widely reported. He pointed his newly built telescope (which wasn't exceptional by any means) towards the planet and frustratingly saw nothing of these. Unknowingly to him at the time, Saturn's position around the Sun had changed since Hevelius'reported observations, and the rings were now displayed as side-view making them practically invisible to an observer on Earth. Nonetheless, this situation proved serendipitous for Huygens, as it resulted in a dimming of the rings' brightness, thereby facilitating the identification of any sizable moon orbiting Saturn.

Perplexed by the missing ansae, the Dutch astronomer continued to observe Saturn and spotted a 'star' nearby, at 3 min of arc away. Acting on intuition, he opted to observe it across several consecutive nights, and by meticulously tracking its movement, Huygens swiftly discerned that it was not a star but rather, a moon. Excitedly, he continued tracking its motion and eventually calculated the time it took for the moon to orbit around Saturn at 16 days. Wasting no time in communicating his observations, Saturn's new moon captivated the public and Huygens became famous overnight. Huygens exhibited keen intellect and remarkable intuition. Besides his discovery of Titan, he was also the first to comprehend that Saturn's enigmatic ansae constituted a colossal ring encircling the planet despite never witnessing the ansae himself, and only relying on drawings derived from earlier observations by Hevelius and his contemporaries.

Titan became the sixth known moon in our Solar System although it wasn't referred to as Titan at the time. Instead, Huygens decided to name the moon Saturni Luna, which means Saturn's moon in Latin. Very soon Saturn's new moon proved exceptional as astronomers recognized it to be very big. With a difference in radius between Ganymede and Titan at only 60 km and the presence of a thick atmosphere on Titan (unknown at the time) which had the effect of extending the moon's apparent size, Titan was erroneously thought to be the largest moon in our Solar System.

It didn't take too long for the prolific Italian astronomer Giovanni Domenico Cassini to start discovering new companions to Saturni Luna. Iapetus was first observed in 1671 and Rhea in 1672. With the existence of these new moons, the designation Saturni Luna became inadequate, prompting a change in the nomenclature of the moons, and similarly to what was employed at the time in the Jovian system, Saturn's moons were ranked by the distance to their planet. Thus, Saturni Luna became Saturn II. Rhea, closer to Saturn, was named Saturn I, while Iapetus, further out, became Saturn III. In 1684, Cassini, continuing his diligent observations, discovered two more moons: Dione and Tethys. With both moons closer to Saturn than Rhea, Tethys now became Saturn I, Dione became Saturn II, Rhea (previously Saturn I) was changed to Saturn III, Titan (previously Saturn II) was changed to Saturn IV, and Iapetus (previously Saturn III) was changed to Saturn V.

Of course, it was only a matter of time before new moons would be unveiled, forcing these names to be changed once more. This occurred a hundred years later when, in 1789, the British-German astronomer William Hershel discovered the smaller moons Mimas and Enceladus lying even closer to Saturn than Tethys. Mimas therefore became Saturn I, Enceladus became Saturn II, Tethys jumped to

Saturn III, Dione changed to Saturn IV, Rhea moved to Saturn V, Titan became Saturn VI, and Iapetus Saturn VII.

Confused? By that time, many people were. Navigating through journals and notes from past observations became increasingly challenging for astronomers, as they had to meticulously track the name changes over the years to correctly identify the planetary objects being referenced.

To address this issue, a new nomenclature was introduced, whereby the names of the moons would become permanent, regardless of any future discoveries of new moons. Under this updated naming convention, Saturn VI would retain its designation as Saturn VI regardless of any additional moons discovered between it and Saturn. Such a solution, however, rendered the naming of the planets rather useless and confusing. It is maybe with this in mind that in 1847, Hershel's son, John, proposed new names for Saturn's moons based on Greek mythology. Accepted by the scientific community without much resistance, the moons were quickly renamed, and thus, Saturn VI became Titan, an appropriate name since it was still thought to be the biggest moon of the Solar System.

The Atmosphere

As with many satellites in our Solar System, Titan remained a mystery for hundreds of years. Other than observing its distinctive orange hue, which set it apart from other moons, little else could be discerned about Titan due to the technological constraints of the era. It was left to speculation as to what Titan might be like. Fast forward to 1907, the moon was put under the spotlight when Spanish astronomer Josep Solà claimed to have observed limb darkening on Titan. This occurs when a planetary body appears darker at its edge or limb, inferring the presence of an atmosphere. Solà's claims seemed dubious though, as he had already made the same announcement for the Galilean moons as well, which possess no such atmosphere. Nevertheless, astronomers' curiosity was aroused. In 1925, Sir James Jeans, a British astronomer, delved into the theoretical study of escape processes in planetary atmospheres. The study included Titan alongside the Galilean moons, revealing that despite its modest gravitational force compared to terrestrial planets like Earth, Titan could theoretically maintain an atmosphere if its surface temperatures were extremely cold, ranging from 60 to 100 K (-213 to -173 °C).

Within this temperature range, gases with a molecular weight of 16 or greater would be unable to attain escape velocity, consequently leading to the formation of an atmosphere around the moon. Such gases included ammonia, argon, neon, nitrogen, and methane.

Furthermore, Sir Jeans predicted that if an atmosphere did indeed exist on Titan, it would be mainly composed of methane, argon, and neon. With methane being more easily detectable through an infrared spectrum than argon and neon which they have weaker absorption bands, he suggested that methane would be the first gas to be detected on Titan once technological advancements permitted such observations in the ensuing decades.

To appreciate Sir Jeans'remarkable foresight, it is necessary to take a step back and comprehend the conditions necessary for the formation of atmospheres on minor planetary bodies.

In Chap. 2, we detailed how, during the formation of our Solar System, most gaseous elements near the Sun couldn't condense because of its radiant energy and the influence of the solar wind. Water, methane, nitrogen, and other volatiles were pushed away of the inner Solar System, providing an explanation as to why the inner planets—Mercury, Venus, Earth, and Mars—primarily consist of rocks and metals.

The water and other volatile compounds that currently exist on the inner planets were not present when they formed but were instead deposited later by comets and ice-rich asteroids.

Upon acquiring the newly introduced volatile compounds, the inner planets developed surface oceans and were shrouded by dense atmospheres, believed to be predominantly composed of nitrogen gas. Thankfully for us, Earth's magnetosphere, as well as geological and biological activity present on our planet, helped sustain the oceans and atmosphere through billions of years. Mars and Venus were less unlucky. Nitrogen and other gases in the early Martian atmosphere were removed by the solar radiation due to the lack of a protective magnetosphere, while Venus suffered the complete opposite, as too much volcanic activity pumped considerable amounts of carbon dioxide and sulfur into the atmosphere, heating it up like a pressure cooker. We review the fate of Venusian and Martian oceans and atmospheres in more detail in the appendices of this book.

As we move outward to the outer planets, the positioning of Jupiter appears to hinder the development of atmospheres on its moons. Indeed, beyond 5 AU, we cross the frost line, where temperatures plummet enough for water to condense into ice, constituting a substantial portion of the composition of icy moons. However, it is still 'too hot' for the condensation of other volatiles such as nitrogen or methane to occur. With no apparent mechanism to deliver these volatiles to the moons, none of the Galilean moons have a substantial atmosphere. Europa, and Ganymede have exospheres, while Callisto's tenuous atmosphere is so exceedingly sparse, it is negligible.

The conditions change once we go further out from the Sun. At Saturn, water once more emerges as a substantial constituent of planetary bodies, however, the temperatures found at this location are just about right to allow ammonia, nitrogen, neon, argon, and methane to become an integral part of the moons' composition, contingent upon each moon's unique conditions.

In Titan's case, ammonia ice served as foundational components from the outset, amalgamating with water ice to shape the moon's icy crust. Subsequently, nitrogen molecules from the ammonia ice underwent conversion into gas, facilitated by various processes like photolysis or exposure to thermal sources. Consequently, a nitrogen-rich atmosphere was established, with nitrogen constituting over 95% of its composition, while argon, methane, and other trace gases account for the remainder, some of which contribute to the formation of dense organic smog. Remarkably, methane and ethane are both found in gas and liquid states, creating cycles of rain and clouds analogous to Earth's water cycle. Despite temperatures averaging 94 K (−179 °C), we see lakes and seas covering roughly 2% of the moon's surface. Introduce both simple and complex organic compounds into this blend of hydrocarbons, and you'll find a fervent community of astrobiologists eager to dispatch robotic probes to its surface.

Had the Saturnian system been situated slightly farther from the Sun, the frigid temperatures would have caused methane and nitrogen to freeze solid onto the ground, and there would be no lakes and seas on Titan's surface.

Astronomers, known for their patience, had to wait two decades after Sir Jeans'prediction before they could perform spectroscopic analysis of Titan's light. This task fell to astronomer Gerard P. Kuiper, who, after spectroscopically observing the ten largest satellites of the Solar System and the dwarf planet Pluto, discovered that only Titan exhibited evidence of an atmosphere. Thus, Sir Jeans'assertion was vindicated.

Moreover, Kuiper identified the methane absorption in Titan's spectra, in line with Sir Jeans' prediction. This revelation had a significant impact on the scientific community, prompting more resources to be allocated to the observation of the moon and its atmosphere. However, due to the constraints of the technology available at the time, minimal progress was made over the following decades.

The situation shifted in 1973 and 1975 with fresh data from ground-based observations revealing dense concentrations of particles in the upper reaches of Titan's atmosphere. This prompted scientists to suggest the existence of hazy clouds formed by condensed methane and intricate organic compounds, including oily droplets resulting from the methane breakdown under UV light and subsequent recombination into polymers. The notion of a moon boasting a dense atmosphere rich in organic compounds was radical. With scientists eager to know more, Titan swiftly became a focal point in planetary science research.

In fact, the moon became of such paramount importance to planetary scientists, that it went straight to the top of the list of the planetary objects to be studied for the yet-to-be-launched Voyager 1 and 2 probes. As such, Voyager 1 was deliberately launched on a trajectory to provide the optimum flyby of Titan, and if it had failed its mission, Voyager 2, arriving a few months later, would have been requested to do the flyby instead, preventing it from visiting the planets Uranus and Neptune; two major planets and their satellite systems of which we knew very little were judged less important than Titan.

Preceding the arrival of the Voyagers, the intrepid Pioneer 11 reached the Saturnian system by September 1979, marking the first spacecraft to achieve this feat; this mission will be examined in more detail in the subsequent chapter focusing on Enceladus. Titan was already listed as a priority target in addition to the giant planet and its rings, and the spacecraft's trajectory took it as close as possible to the moon at 360,000 km.

Upon reaching Saturn, Pioneer 11 captured the initial images of Titan; five in total. Upon examination of the blurred, obscured disk, see Fig. 7.1, scientists swiftly realized that the dense atmosphere would pose a significant obstacle to the study of

Fig. 7.1 The first image of Titan by Pioneer 11 taken on September 2, 1979, from 360,000 km. Comprising five images, the quality of the composite is constrained by the limited capabilities of the Pioneer imaging system and the rudimentary state of telecommunications during the Titan encounter. (Image courtesy of NASA)

the moon itself. Pioneer 11 did unveil Titan as an extremely frigid environment, with an average temperature hovering around 94 K (−179 °C), effectively eliminating any possibility of life as we understand it emerging there.

Voyager 1 and Cassini-Huygens

One year later, planetary scientists were elated as Voyager 1 navigated the Saturnian system with success, culminating in its close encounter with Titan at a mere distance of 4394 km. It was a brief encounter, lasting only a few hours. During this flyby, it was revealed that Titan's diameter was slightly smaller than that of Ganymede, reinstating Ganymede as the largest moon in the Solar System. Voyager 1 also enabled scientists to determine Titan's density, which was found to be comparable to Ganymede and Callisto; approximately half water and half rock. At that time, the concept of subsurface oceans beneath the Galilean moons was beginning to gain consideration, although it was still premature for scientists to recognize the potential existence of such an ocean on Titan as well.

The imaging data transmitted by Voyager 1 proved to be profoundly disappointing as Titan's dense atmosphere completely obscured the surface, rendering any features undetectable. Nonetheless, the composition of Titan's atmosphere was fully revealed, consisting primarily of nitrogen (98% in the upper atmosphere), methane (between 2% and 8%), along with trace amounts of hydrogen, carbon monoxide, carbon dioxide, as well as various organic gases including ethane, propane, acetylene, ethylene, hydrogen cyanide, diacetylene, methyl acetylene, cyanoacetylene, and cyanogen.

The presence of nitrogen in Titan's atmosphere suggests a geologically dynamic history. When frozen ammonia (NH_3) is heated, it vaporizes into nitrogen, which is subsequently released into the atmosphere through cryovolcanic activity. Sunlight also plays a role as it breaks down surface ammonia ice into hydrogen and nitrogen, which then bond to form the nitrogen molecule N_2.

Voyager 1's discovery of methane in the atmosphere, while anticipated by Sir James Jeans, remains intriguing due to its breakdown by sunlight into hydrogen, resulting in a relatively short lifespan of tens of millions of years. Thus, methane suggests the involvement of replenishing mechanisms with geological processes emerging as the most probable source, although the possibility of biological origins cannot be entirely discounted—it is worth noting that only 5% of methane on Earth is generated by geological processes.

Researchers further discovered that, when methane is broken down, carbon atoms reassemble to create intricate organic molecules, some of which condense in the atmosphere and settle on the surface, forming thick oily substances known as tholins. The notion of organic molecules precipitating onto the moon's surface captured the imagination of both the scientific community and the public. Titan stood out as unparalleled in our Solar System, sparking a strong desire for further exploration.

Following the successful flyby of Titan, NASA and ESA teamed up to develop a flagship mission—similar in scope to the Galileo mission—focused on exploring the Saturnian system: the Cassini-Huygens orbiter and lander. Alongside the planet and its rings, Titan emerged as the primary scientific target. ESA took charge of designing and constructing the Huygens lander, the ground-breaking mission that would traverse through the moon's dense atmosphere and land on its surface, marking a historic first for the outer Solar System.

Prior to Cassini-Huygens' arrival in 2004, planetary scientists saw with great interest the suggestion that the Galilean moons could host subsurface oceans and considered if this would be possible for the moons of Saturn as well. Using theoretical models founded on the thermal and mechanical characteristics of ices and silicates, they also posited the potential existence of a subsurface ocean composed of liquid water on Titan.

The involvement of ammonia (NH_3), a readily available resource around Saturn, played a vital role in their models as the freezing point of ammonia-rich water decreases substantially due to ammonia's low freezing point at 196 K (−77 °C). The models suggested that depending on the concentration of ammonia and other volatiles present within the primordial liquid layer, the physical conditions might be right for the formation of a global subsurface ocean of water and ammonia situated 350 km under the icy shell. This was an exciting possibility. All that was required now was to prove it. Luckily, the Cassini-Huygens mission was on its way.

Throughout its primary mission spanning from 2004 to 2017, the Cassini orbiter executed over a hundred flybys of Titan, meticulously mapping the colossal moon, while the Huygens lander achieved a successful touchdown and conducted on-site surface analysis. The discoveries made by these daring voyagers are nothing short of extraordinary: methane and ethane lakes and shallow seas adorn the northern hemisphere, while hydrocarbon-rich dunes blanket the equator. These diverse environments have opened new avenues in the quest for extraterrestrial life, although given surface temperatures plunging to 95 K (-178 °C), any potential life would truly be alien, bearing no resemblance to life on Earth.

As fascinating as the surface proved to be, planetary scientists were also keen to find the evidence for a water environment on Titan deep below the frozen crust. Could a subsurface ocean exist? The models on Titan's thermal evolution suggested the possibility of a subsurface ocean, as radiogenic and tidal heat seemed sufficient to melt the interior ices. Luckily it didn't take long for the first tantalizing clues of a subsurface ocean to appear.

The Evidence of a Subsurface Ocean

On January 14, 2005, the Huygens probe performed the most distant and daring landing of any human-made craft. After a 2 hours descent through Titan's atmosphere, it touched down on a fluvial basin covered with organic-rich material and dotted with pebbles most likely made up of hydrocarbon coated water-ice, see Fig. 7.2. An array of scientific instruments had captured data all the way down, although things didn't go entirely to plan as a misfortunate software error prevented the transmission of half of the images taken by the probe as well as the loss of all the data on wind speed and Doppler radio measurements.

Amid the scientific instruments carried by Huygens was the 'Permittivity, Waves, and Altimetry' sensor, also known as PWA, which was part of the Huygens Atmosphere Structure Instrument. During the descent, PWA measured the ambient electric field in Titan's atmosphere. It achieved this by detecting extremely low frequencies (ELF), which are characterized by radio waves oscillating at a very slow rate, specifically 36 times per second. On our planet, ELFs, often referred to as Schumann resonance, are produced by thousands of lightning bolts taking place each minute within the atmosphere. Earth was the only known planetary object in our Solar System where ELFs had been detected, and since lightning was thought to be prevalent in the moon's atmosphere—a microphone, a first for an interplanetary mission, had been installed on the lander to listen for thunder—scientists assumed that ELFs could be detected on Titan as well.

Although lightning wasn't detected, the PWA data revealed a narrow-band signal at about 36 Hz during Huygens' descent to Titan. A clear indication that ELFs had been produced despite the absence of lighting. This was a fortunate discovery as ELFs can penetrate deep into the ground and continue until they hit a layer of conductive material which then reflects them back out. On Earth, the ground is highly

Fig. 7.2 Titan's surface as
seen by Huygens. This is
the first image from a
planetary surface within
the outer Solar System.
Globules, probably made
of water-ice, ten to fifteen
centimeters in size lie
above a darker, finer-
grained substrate in
variable spatial
distribution. (Image
courtesy of ESA/NASA)

conductivity and ELFs bounce back as soon as they hit the surface. Titan's surface,
on the other hand, is characterized low conductivity allowing ELFs to pass straight
through the top layer until something reflects them back. It seemed that the PWA
had detected ELFs bouncing back from the surface.

After carefully examination, scientists at ESA determined that TItan's ELFs are
trapped in a 'cavity' between the moon's ionosphere, between 40 and 140 km in
altitude, and a lower boundary, some 55–80 km below Titan's surface, where a conductive layer is present. The scientists involved with the PWA instrument suggested
that either a thick layer of liquid water or salts mixed with ammonia could generate
the conductivity required to produce the observed pattern. And thus, without expecting it, the Huygens team seemed to have discovered a first hint at Titan's subsurface
ocean. Not everyone was convinced of this interpretation though, and still to this

day, the Schumann resonance detected by Huygens is seen, at best, as weak evidence in support of a subsurface ocean.

Luckily, stronger evidence for a subsurface ocean was on its way. In 2021, new data came from a team led by Luciano Iess from Sapienza University in Rome. Iess' team tracked with great accuracy Titan's shape as it orbited Saturn every 16 days.

As Titan follows a slightly eccentric orbit, at its closest to Saturn (periapsis), the planet's substantial mass stretches the moon, elongating it like a rugby ball, while at the farthest point (apoapsis), the moon reverts to a more natural spheroid shape. Since scientists couldn't precisely measure the moon's shape due to its thick atmosphere, an alternative solution was found as minute fluctuations in the spacecraft's velocity due to the moon's gravity could be measured. To track the spacecraft as accurately as possible, Cassini's Radio Science Subsystem sent radio signals at 33,000 and 140,000 bits per second which NASA's Deep Space Network pick up back on Earth. By precisely measuring when these signals arrived, the team could determine the spacecraft's velocity at a given time, and thereby interpret the effect of gravity from a nearby object. By doing so, the scientists had detected the gravitational effect driven by the variation of Titan's shape and mass along its orbit around Saturn.[2]

By analyzing the radio signals, Luciano Iess and his team discovered that Titan experiences tides reaching heights of up to 10 m as it orbits Saturn. Physical models have indicated that if Titan's interior were entirely frozen solid, such tides would be significantly smaller, at only 1 m. The observed tides on Titan can only be accounted for if a thick icy crust sits atop a flexible mantle of liquid located deep below the surface, finally given planetary scientists what they were looking for: the proof of Titan's subsurface ocean.

And so, following Callisto, Europa, Ganymede, and Enceladus (see next chapter), Titan emerges as the fifth confirmed ocean world within our Solar System. With its surface hosting complex organic chemistry, Titan ascends to an even higher level of significance as a planetary body.

It is worth pointing out here that a paper appearing in 2008 reported apparent shifts in surface features, suggesting a variability of Titan's rotation period by about 0.36°, which in turn would hint at a subsurface ocean. This was later disproved as an artifact of early engineering software used to analyze the radar data (Fig. 7.3).

The Ocean and Its Habitability

What can we infer from these findings? Titan has a global subsurface ocean a few 100 km thick lying under a de-coupled icy shell, meaning that the icy crust slides on the liquid mantle. This shell, mostly made of Ice Ih and some ammonia, has an

---

[2] Gravity cannot be directly measured by a spacecraft due to its inherent motion in space. Only by interpreting the variations in the radio waves emitted by the spacecraft can we infer the gravity effect on a spacecraft.

TITAN

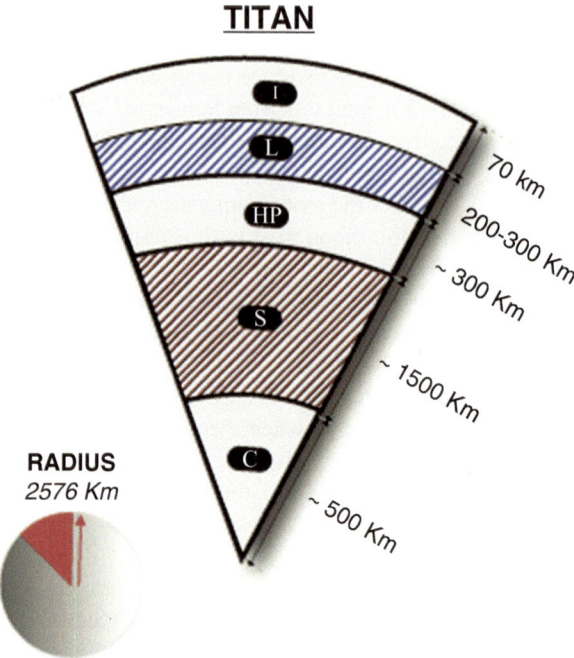

Fig. 7.3 Diagram showing the interior of Titan where the subsurface ocean lies between two layers or thick ices. The thickness of the mantles is not well known. *I* Ice, *L* Liquid, *S* Silicate mantle, *C* Core. Diagram is not to scale.

average thickness of around 70 km, although recent studies have instead implied that in some areas, it could be as thin as 10–50 km. Gravity and topography data collected by Cassini over a decade have found that the subsurface ocean is of high density, like the Dead Sea back on Earth. This suggests that the ocean is charged with salts, most likely sulfur, sodium, and potassium. Ammonia is also thought to exist in significant amounts, enabling the ocean to remain in liquid form at extremely low temperatures.

This subsurface ocean rests on a high-pressure layer of solid ice, most likely Ice VI, although the exact phase of ice is currently unknown—see Chap. 4 for further details on the ice phases. This icy mantle then rests on a rocky mantle a few thousand kilometers in diameter. So, much like the oceans of Callisto and Ganymede, Titan's subsurface ocean is sandwiched between two thick layers of ice, the icy crust at the top and the HP icy mantle at the bottom sealing it from the rocky material below.

Nevertheless, as we saw in our chapter on Ganymede, this doesn't necessarily imply that the ocean is totally isolated. In theory, warm bubbles of ice charged with minerals and chemicals originating from the silicate mantle could rise within the HP ice, and reach the subsurface ocean. It is a very remote possibility, but a possibility, nevertheless.

It is also worth noting that the Cassini orbiter detected slight traces of argon-40 in Titan's atmosphere, which is intriguing since this isotope of argon is the result of the decay in potassium-40, an element found in rocks. This might be suggestive of past water-rock interaction, although where this interaction took place is unknown.

Considering Titan's surface abundant in organics, is it possible for any interactions to occur between the subsurface ocean? The thickness of the icy crust is a severe impediment, although one of the most recent finds has been the discovery of cryovolcanoes on the moon's surface. Researchers analyzing topographical data discovered a region called Sotra Facula, where three large volcanic cones, 1–1.5 km tall, were found to have deep pits. It is likely that these could have been releasing materials from the subsurface into Titan's atmosphere in the past, potentially replenishing it with methane.

Also, evidence of resurfacing has been detected in certain places, suggestive of past geological activities most likely due to cryovolcanoes. While these findings shed light on potential heat exchanges within the icy crust, and could indicate subsurface material reaching the surface, the notion of organic-rich compounds from the surface being transported into the icy crust and subsequently reaching the subsurface ocean deep underground appears highly unlikely.

Titan's subterranean sea is undeniably alien; a frigid amalgamation of water, ammonia, and salts, scarcely interacting with minerals and organic matter. The probability of life as we know it emerging in such an environment is extremely remote. Alas, given that the ocean remains sealed under a substantial icy mantle, we are unlikely to ever know what lies within it.

There is one subsurface ocean in our Solar System, though, where this is not the case. Hosting deep cracks within its icy crust and venting seawater directly into space, a tiny moon orbiting Saturn supports the most well understood ocean after Earth's surface oceans. It is now time for us to visit the most promising of the confirmed ocean worlds in our Solar System: Enceladus.

Chapter 8
Enceladus

After our visit to the giant moon Titan in the previous chapter, the small-scale moon of Enceladus is striking. Ten times smaller than Titan, and almost one hundred times less massive, this is a planetary object on a very different scale than the giant moons we have been reviewing so far. Another arresting feature of the tiny moon is how bright its surface is. With an albedo of 1.34, Enceladus his the most reflective object in our Solar System. This suggests a very young surface.

Perhaps most astounding of all, this tiny moon harbors giant fissures located within the south pole which vent seawater into space. These vents in turn feed the vast nebulous E ring surrounding Saturn and Enceladus' neighboring moons. Despite its tiny size, Enceladus demonstrates an unparalleled capability to exceed expectations.

Flying Through the E Ring

Enceladus was discovered with Mimas more than a 100 years after Titan, in 1789, by William Herschel, a British astronomer of German origin, better known for his discovery of planet Uranus. By then four additional moons had already been discovered orbiting Saturn: Iapetus (1671), Rhea (1672), Dione (1684), and Tethys (1684).

Initially, Enceladus was named Saturn II following the naming convention used at the time when the moon was discovered. (We detailed in chapter seven how this confusing naming convention came to be). It was Herschel's son, John, who later suggested mythological names for all of Saturn's known moons, giving us Enceladus.

Once discovered, astronomers would closely follow the orbits of the Saturnian moons, and in the case of Enceladus and Dione, they quickly realised that the moons were locked in a 2:1 orbital resonance with each other. On other words, while Enceladus makes two revolutions around Saturn, Dione makes precisely only one. At the time, astronomers looked at these resonances more as mere orbital curiosities

B. Henin, *Exploring the Ocean Worlds of Our Solar System*, Astronomers' Universe,
https://doi.org/10.1007/978-3-031-62953-2_8

driven by the peculiarities of space mechanics than anything worth investigating. At the time, the orbital predictions of Saturn's moons were fraught with errors into the tens of thousands of kilometers. But it didn't matter. Apart from Titan, the smaller-sized moons of Saturn were thought to be inert balls of ice and rocks and were mostly ignored. In fact, before the arrival of space probes, knowledge about the orbits of Saturn's moons was still very poor, with most measurements dating back from the 1920s and 1930s.

Shortly after Earth-based photometric instruments became capable of providing noteworthy measurements of the Saturnian moons, astronomers started to pay attention to them again. They noted differences in brightness between the leading and trailing hemispheres of Mimas, Enceladus, Tethys, Rhea, and Dione. Interestingly though, the leading sides of Tethys, Dione, and Rhea were brighter than the trailing side, while the opposite was true for Mimas and Enceladus. No explanation could be given at the time for these differences. Furthermore, Enceladus and Thetys displayed high albedos; noteworthy, but not necessarily extraordinary, for icy bodies. Regardless, technological limitations hindered any further studies of these moons and their characterization.

All this changed in 1966 when Walter Feibelman, an astronomer at the University of Pittsburgh, spotted what seemed to him as a faint whitish halo located within the orbits of the icy moons. While the evidence suggested a new ring system surrounding Saturn, Feibelman's interpretation remained controversial due to the faint and diffuse nature of the potential ring. At the time, only three distinct ring systems were recognized: the A (outermost), B (middle), and C (innermost) rings.

Feibelman was convinced that he had found a new ring and named it the E ring. Follow-up observations merely suggested at a diffuse ring between the orbits of Mimas and Titan, with a curious thickening around Enceladus'orbit. Due to the whitish aspect of the E ring and the icy nature of Enceladus, could this seemingly inert moon be the source of the hypothetical ring? Skepticism was high. Existing geological knowledge couldn't explain how an inactive moon could eject material into space. Asteroid impacts offered the only plausible explanation, though the lack of similar rings around neighboring moons raised doubts. Astronomers were puzzled and placed their hopes on the space age and its latest marvels: space probes. Pioneer 11 would be the first to visit.

The 1970s saw a resurgence in the study of the Saturnian system in preparation for trail-blazing space probes, such as Pioneer 11 and the Voyagers. When Pioneer 11 was launched in 1973, mission planners had less than 6 years to work out and agree on the precise trajectory the spacecraft would take. By 1975, more accurate measurements of the moons' orbits and the development of a modern theory of motion bolstered by new mathematical tools gave researchers a high level of precision and confidence in their predictions.

Nevertheless, agreeing on the final course Pioneer 11 would take proved much harder than anyone had anticipated. The initial path advocated by the Principle Investigators—the mission scientists responsible for the quality and direction of the scientific research—was to pass between the rings and Saturn itself, a supposedly debris-free space close to the planet's surface. Such a pass would provide invaluable information on Saturn's magnetic field, its radiation belt, and the potential interaction between these and the rings. This would be known as the "inside option." Some members of the Pioneer 11 team also contemplated sending the space probe through the Cassini division, the biggest of the ring gaps separating rings B and A. Given that this 'gap' would be later found to be populated by particles similar in size to the C ring, albeit at lesser density, it is fortunate that this path wasn't chosen (Fig. 8.1).

Pioneer 11, a twin of Pioneer 10 which we referred to in the previous chapters on the moons of Jupiter, was primarily conceived to investigate particle and field science and had limited capabilities for imaging science. The decision to opt for the "inside option" was primarily aimed at maximizing the scientific output from the particle and field science instruments. Given that Pioneer 11 was not slated to explore any further planets beyond Saturn, this approach made logical sense. It allowed the mission team to pursue a more daring maneuver, ensuring a significant science return to the mission, despite its inherent risks.

The Pioneer 11 mission was not operating in isolation though. As soon as the U.S. Congress agreed to fund the "Mariner Jupiter-Saturn" mission in 1972—later renamed the Voyager program—Pioneer 11's fate was sealed, as it now had a new mission: finding a safest path through the Saturnian system for the Voyagers to traverse.

This new approach would prove crucial for the exploration of the outer planets. Although the Congress had initially agreed to fund the Mariner Jupiter-Saturn mission which, as its name indicates, was expected to end in the Saturnian system, scientists secretly hoped for further funding, allowing for the continuation of the mission after Saturn and the visit of Uranus and Neptune.

Initially, this concept was introduced within the Grand Tour program. However, the Grand Tour initiative was terminated in December 1971, primarily due to escalating costs associated with the newly sanctioned Space Shuttle program. Consequently, in the absence of extended Voyager missions, there were no alternative plans in place to explore the two farthest planets in our Solar System.

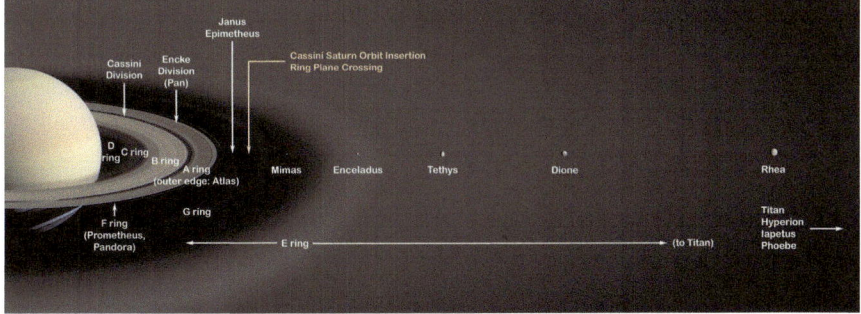

Fig. 8.1 This illustration depicts Saturn's rings and prominent icy moons. Saturn's ring system forms a vast and intricate structure. In terms of scale, the entire expanse of the ring system would not even fit in the space between Earth and the Moon. The seven main rings are labeled in the order in which they were discovered. From the planet outward, they are D, C, B, A, F, G, and E. (Image courtesy of NASA/JPL)

If Congress would agree to extend the Mariner Jupiter-Saturn program, one of the Voyagers could use Saturn as a slingshot, placing it on a direct trajectory to Uranus followed by Neptune. This was due to a particular alignment of planets that only occurred once every 175 years. It was an opportunity not to be missed, and ultimately, the renamed Voyager 2 would be chosen to do such a tour.

Regarding Voyager 1, it had been determined that the spacecraft would conduct a close flyby of Titan, a pivotal objective of the entire mission; see the previous chapter on Titan for further insights. This necessitated an upward trajectory out of the ecliptic plane. Consequently, even with a mission extension, Voyager 1 would not be able to visit Uranus and Neptune.

Also, as trajectories taken by the Voyagers within the Saturnian system were being calculated, it was made apparent that due to the constraints of orbital mechanics, a mission extension to the icy giants would require Voyager 2 to come close to the edge of the A ring where a section of the E ring was thought to be located. Voyager 1 would stay far away from the rings on its way towards Titan. The trajectory selected for Voyager 2, known as the "outside track," became a significant concern for the mission planners.

Astronomers knew that each ring was composed of tens of thousands of individual ringlets, but their exact densities was unknown, as was the dust concentration within the gaps themselves.

The Voyager team encountered a challenge: determining the potential impact of particles on the spacecraft. Engineers calculated that particles smaller than a millimeter wouldn't pose a significant threat, allowing the spacecraft to navigate through unharmed. However, larger particles posed a considerable risk. Studies indicated that once particles exceeded 1 cm in size, their distribution became more dispersed, reducing the likelihood of impact. However, the average size of particles in the E ring remained unknown. Adding to the complexity, a few years after the apparent (and still unconfirmed) discovery of the E ring, another very faint ring seemed to have been found. This time located between the C ring and the planet. It would later be named the D ring. It was therefore entirely plausible that other diffuse rings could lie undetected within Saturn's orbit and be in the path taken by the oncoming spacecraft. The only way to test if the outside track would be safe for Voyager 2 was to take advantage of Pioneer 11's early arrival and test the passage. If the spacecraft passed unscathed, then the passage would be clear for Voyager 2 to follow.

There was a drawback though; if both Pioneer 11 and Voyager 2 were to follow the outside track successfully, neither would get close enough to Saturn to ensure high-quality measurements of the planet. Instead, if Pioneer 11 would suffer any form of damage or even get destroyed as it forced its way through the E ring, then the outside track would be deemed unsuitable for Voyager 2, preventing it from continuing to the outer planets. In this scenario, Voyager 2, free from a predetermined outward trajectory, could explore the Saturnian system along a path more conducive to studying the giant planet itself.

Time was ticking though, and the decision to send Pioneer 11 on the inside or outside track had to be made by the middle of 1978 at the latest; by then, Voyager 2 had already been launched and was well on its way to the Jovian system. With the Pioneers and the Voyagers built by different teams, Pioneer 11 was built by the

Ames Research Center and Voyager 2 by the Jet Propulsion Laboratory, reaching a consensus for the ultimate path taken by Pioneer 11 was difficult.

Ames aimed to maximize the scientific output from the instruments aboard their spacecraft, the first to venture to Saturn. They advocated strongly for their space-craft to take the inside track. Conversely, JPL advocated for Pioneer 11 to take the outside track in anticipation of Voyager 2's arrival. Ultimately, the decision would fall on the director of the Planetary Program at NASA, held at the time by a brilliant engineer and manager named A. Thomas Young. After carefully examining both options, he sided with strategy supporting the long-term exploration of our Solar System and thus, instructed that Pioneer 11 make the necessary course corrections to place itself on the outside track, paving the way for Voyager 2.

Therefore, on September 1, 1979, as Voyager 2 was en route to the Saturnian system, Pioneer 11 traversed the plane of the E ring, at a velocity close to 114,000 km per hour. This hair-raising moment lasted just a few seconds, after which the space-craft successfully carried on its course. Unharmed, Pioneer 11 left the door wide open for Voyager 2. Young's decision proved to be the correct one, leading scientists to seriously consider the feasibility of a spacecraft visiting Uranus and Neptune.

Ironically, Pioneer 11 couldn't confirm the presence of the E ring due to techni-cal limitations of the instruments it carried, but it did nevertheless return valuable science data on the giant planet, detected two new rings and two new moons, and also studied Titan's atmosphere. However, due to the constraints inherent in the outer track trajectory, the space probe whizzed by Enceladus at 222,027 km, too far to take a picture. And so, as with most of the moons of our outer Solar System, it would be the Voyagers that would bring Enceladus to life.

The Voyagers' Encounter

Voyager 1 was the first of the Voyagers to enter the Saturnian system, and on November 12, 1980, the spacecraft sent images of new worlds: Mimas, Rhea, and Tethys. At its closest approach to Enceladus, approximately 202,000 km away, Voyager 2 didn't get much closer than Pioneer 11. However, equipped with signifi-cantly more advanced imaging instruments, the spacecraft captured the first-ever image of Enceladus. This image caught everyone off guard. Contrary to expecta-tions, large regions of the surface appeared devoid of craters, while extensive grooves and cracks were evident. The remarkably youthful surface of Enceladus prompted comparisons to Europa or Ganymede by planetary scientists.

After Voyager 1's visit, and before Voyager 2's arrival, scientists were grappling with the mystery of how such a diminutive moon could possess adequate energy to drive what appeared to be recent geological processes. Could tidal heating newly discovered in the Jovian system be at play here as well?

Voyager 1 also validated the presence of the E ring and affirmed Enceladus's position within its most concentrated region. The moon's proximity to the E ring was evident, although the precise mechanisms responsible for its formation remained elusive to scientists.

Fig. 8.2 A closeup view
of the surface of Enceladus
obtained on Aug. 25, 1981,
when Voyager 2 was
112,000 km away. Notice
the large surface areas
devoid of craters and the
young, grooved terrains
suggestive of interior
melting. The largest crater
visible is about 35 km
across. (Image courtesy of
NASA/JPL)

On June 5, 1981, Voyager 2 whizzed by Enceladus at 87,010 km, returning the first high-resolution images of the surface; more specifically, its trailing hemisphere. These were mesmerizing (Fig. 8.2).

Fault lines, ridges, valleys, plains, and long straight grooves were visible, depicting a highly active world; Enceladus was found to be the most active of all Saturn's moons.

The moon's surface appeared to undergo cycles of melting and refreezing, exhibiting signs of both stretching and compression. Furthermore, its remarkably high albedo, the brightest surface reflection in the Solar System, suggested that fresh ice particles were periodically deposited on the surface, possibly expelled by cryovolcanoes. However, no current activity was detected. Whatever processes were at play in reshaping the moon's icy crust seemed to originate from deep within its interior.

The scientists were taken aback. They were still processing the revelations from the Voyagers' exploration of Europa, and now had to contemplate that another geologically active moon within our Solar System had been found.

However, not everything went as expected. Voyager 2's predetermined trajectory had it pass behind Saturn, blocking any contact with Earth for 2 hours and 20 minutes. This wasn't a problem as all the data captured by the spacecraft could be stored by the onboard computer until communications could be re-established, and transmissions resumed. However, it was during this extended blackout that Voyager 2 would make the perilous crossing of the E ring and perform the close flybys of Enceladus and Tethys.

Given the risky nature of the maneuver, both engineers and scientists felt apprehensive when communication with Voyager 2 was lost as it passed behind the ringed Planet.

Fortunately, the spacecraft regained communication after passing behind Saturn, much to everyone's relief. However, upon reviewing the new data, mission engineers quickly noticed an issue: some of the supposedly captured images of the moons appeared entirely black. As more flawed images and data streamed in, the Voyager team quickly deduced the cause of the issue to the dismay of all involved. It appeared that the camera platform had malfunctioned while the spacecraft passed behind Saturn, causing the cameras to veer off course. Somehow, the mechanism supporting the cameras and other instruments had become stuck. Engineers pinpointed the technical glitch to one of the swivels on the mobile platform, occurring 45 min after the spacecraft traversed the E ring. It seemed therefore likely that the probe had collided with an icy particle located farther out in the E ring. Ultimately, the cause of the faulty swivel would never be known.

The impact with the icy particle had two consequences. Firstly, Voyager 2's flybys of Uranus and Neptune would have to be reconfigured to consider this new technical limitation imposed on the spacecraft.

Secondly, and more unfortunately for us, the loss of multiple close-up images of Enceladus meant that the Voyager mission missed an opportunity to observe the jets of water erupting from the moon's south pole. Had this problem not occurred, such a discovery made in 1981 would likely have influenced the conception of the Cassini orbiter, launched a decade later, as the study of these plumes would have become a top priority. Such an outcome might have already reshaped our perspective on life in the Solar System and our place within it. Said differently, a mere speck of ice might have altered the course of history.

A final anecdote from the Voyagers' flyby of the Saturnian system underscores how close we came to discovering the erupting plumes of Enceladus. In early 2017, Ted Stryke, a professor at a U.S. college, re-examined images captured by the wide-angle cameras of Voyager 1 during its encounter with Saturn 35 years earlier. These low-resolution images, primarily used for navigation purposes to pinpoint the spacecraft's exact position, held limited significance for mission scientists. However, leveraging modern imaging tools, Stryke processed the low-resolution images whenever Enceladus appeared in view of the wide-angle camera. Astonishingly, he identified a plume venting from the moon's south pole. The Voyager mission's imaging team couldn't have replicated Stryke's discovery simply because they were unaware of the existence of such plumes and thus never actively searched for them.

Cassini's Arrival

Expanding upon the achievements of the Voyager missions, NASA and ESA embarked on a collaborative effort to undertake the most intricate planetary science mission ever formulated: the Cassini-Huygens mission. The mission was composed of the Cassini orbiter, weighing 6 tons, aimed to meticulously investigate the Saturnian system, and the Huygens lander designated to conduct surface investigations of Titan. The Cassini-Huygens mission would become the most successful planetary mission developed by both agencies thanks to the extensive scientific data amassed through its suite of instruments, including four optical instruments, six fields and particle instruments, and two microwave remote-sensing instruments.

Lifted from Florida in 1997 by a Titan IVB/Centaur launch vehicle under the auspices of the U.S. Air Force, the spacecraft embarked on a seven-year odyssey to reach Saturn, utilizing gravitational assists from Venus, Earth, and Jupiter, while completing two full revolutions around the Sun.

Throughout its primary mission, spanning from July 2004 to 2008, the orbiter's primary emphasis was on scrutinizing the planet itself along with its rings, Titan, and as many of the icy moons as feasible, including Enceladus. The principal objective concerning Enceladus was to investigate further the observations made by Voyager, either confirming or refuting the notion of the moon possessing an atmosphere or undergoing eruptions. However, Enceladus quickly turned out to be more than an intriguing object to study. It became one of the most fascinating places in our Solar System.

Cassini was scheduled to conduct a sequence of prearranged flybys of Enceladus approximately 1 year after the onset of the primary mission. These would bring the orbiter to three exceptionally close approaches of the moon. The first pass, referred as E0, occurred on February 17, 2005, and saw the Cassini orbiter approach Enceladus at a distance of 1264 km. The second pass (E1) brought the spacecraft closer, at 500 km, on March 9, while the final pass (E2) planned on July 3, was to take place at around 1000 km. In addition to these close passes, further observations, referred as non-targeted flybys, were to be performed of the moon at much greater distances (100,000 km or more) (Table 8.1).

Just before the probe's initial flyby in January 2005, the now-famous plume was first observed, thanks to images captured by the Imaging Science Subsystem (ISS). These images revealed a faint plume appearing to originate from the south polar region, slightly brighter than the background noise in the image. Given the exceptional nature of the plume discovery, additional observations were deemed necessary for mission scientists to confirm their interpretations.

A few days prior to the first pass, the spacecraft was positioned in a way to conduct observations during a stellar occultation (a star passing behind the moon) illuminating Enceladus' equator. The ultraviolet imaging spectrograph subsystem (UVIS) onboard Cassini had been specially designed to conduct science during occultations, and scientists were hoping that the UVIS data could finally shed some light on what was occurring around the icy moon. UVIS was a set of telescopes used to measure ultraviolet light from atmospheres, rings, and surfaces. During occultations, it could observe fluctuations of starlight moving behind planetary objects, allowing the characterisation of planetary atmospheres. The observations were made on Enceladus, and as expected, UVIS didn't detect an atmosphere. The moon was too small to retain an atmosphere.

And then came the first flyby, E0. What Cassini observed during this first pass was short of extraordinary. High-resolution images unveiled regions previously surveyed by Voyager 2 and showed that the surface was characterized by multiple fractures. Cassini also verified the existence of water ice on the surface and detected fractures near the south pole containing basic organic compounds and carbon

Table 8.1 Throughout its 13-year mission, Cassini conducted numerous close flybys of Enceladus. The orbit number indicates Cassini's position in its orbit around Saturn. In total, Cassini conducted 23 close flybys.

Orbit number around Saturn	Encounter name	Date	Distance (km)
3	E0	February 17, 2005	1264
4	E1	March 9, 2005	500
11	E2	July 14, 2005	175
61	E3	March 12, 2008	48
80	E4	August 11, 2008	54
88	E5	October 9, 2008	25
91	E6	October 31, 2008	200
120	E7	November 2, 2009	103
121	E8	November 21, 2009	1607
130	E9	April 28, 2010	103
131	E10	May 18, 2010	201
136	E11	August 13, 2010	2554
141	E12	November 30, 2010	48
142	E13	December 21, 2010	50
154	E14	October 1, 2011	99
155	E15	October 19, 2011	1231
156	E16	November 6, 2011	496
163	E17	March 27, 2012	74
164	E18	April 14, 2012	74
165	E19	May 2, 2012	74
223	E20	October 14, 2015	1839
224	E21	October 28, 2015	49
228	E22	December 19, 2015	4999

dioxide. Additionally, magnetometer observations unveiled distinct distortions in Saturn's magnetic field near the moon, indicating its influence. Moreover, water was identified as a significant source of ions for the magnetospheric plasma observed around the moon, suggesting the venting of water vapor into space. Essentially, while the imaging team had initially discovered the particle component of the plume a month earlier, the magnetometer team uncovered the vapor component of the plume.[1]

The observations suggested at a very tenuous exosphere or a plume-like feature on Enceladus. A month later, E1 took Cassini on its second closest approach at 500 km. The orbiter located a source of plasma composed of water ions in the southern region but yielded no further insight into the plume itself.

The final rendezvous with Enceladus was scheduled for July, and with no further flybys slated for the remainder of the prime mission, time was running out fast if the team wanted to investigate the plume further. In a "now-or-never" scenario, the Cassini program manager, prompted by the magnetometer team, agreed to a daring

---

[1] There have been disagreements within the Cassini team as to which instrument was the first to detect Enceladus' plume with some scientists having publicly taken issue with the official storyline.

maneuver for this last encounter. Instead of maintaining the original plan to approach within 1000 km, Cassini would skim the surface at a hair-raising distance of just 175 km. This adjustment was no trivial matter as it required significant course corrections and alterations to Cassini's trajectory around Saturn, as well as additional fuel consumption.

Luckily, in addition to the data that would be returned by this new flyby, another stellar occultation could be performed during the approach, ensuring that UVIS would be able to take additional measurements of the plume.

Hence, when Cassini conducted its closest flyby of Enceladus on July 3, mission scientists awaited the incoming data with heightened anticipation. The data they were sent back surpassed their expectations.

For a start, the third flyby allowed close shots of the southern hemisphere with unprecedented image scales of up to 4 m per pixel. The images revealed that the south pole region, starting at around 55° in latitude south, exhibited numerous features formed by recent geological activity such as ridges and fractures with little crater impacts present suggesting a very young age. Also, the albedo of the southern region was brighter compared to the rest of the moon, indicating that it had recently been covered with fresh ice. Most intriguing was the presence of long fissures, now known as tiger stripes due to their resemblance to claw marks, embedded in 200-m-high ridges and extending for up to 130 km in length. Each stripe ran parallel to each other at an average distance of 40 km. In November 2006, each stripe was named after a city named in the Arabian Nights folk tale followed by the Latin term sulcus (plural, sulci), meaning a groove or a ridge: Alexandria Sulcus, Cairo Sulcus, Baghdad Sulcus, and Damascus Sulcus; the latter two being the most active sulci at present.

It is worth noting here that when Voyager 2 made its closest approach to Enceladus years earlier, it had flown over the northern region, where stripes were non-existent. One could only imagine the consequences if Voyager 2 had flown over the south pole instead and had discovered the tiger stripes in the early eighties.

Particularly notable were the results from the composite infrared spectrometer (CIRS) that measured surface temperatures.

CIRS uncovered that the south polar region of Enceladus was warmer than the moon's equator, which receives the most sunlight energy. The highest temperatures were detected within the tiger stripes themselves, ranging between 114 and 157 K (-159 to -116 °C).

Cassini's instruments also assessed the ice's crystallization state in the southern region, revealing pristine, fresh ice near the tiger stripes. In contrast, radiation-damaged amorphous ice was found farther away, suggesting recent venting of fresh ice from the stripes, possibly within the last 1000 years. Light compounds, such as methane, were also found near the stripes. In addition, results from two instruments that analyzed particles in space, the cosmic dust analyzer (CDA) and the ion neutral mass spectrometer (INMS), showed considerable increase of water particles within the vicinity of the moon, confirming that Enceladus was indeed the primary source for the E ring.

Furthermore, the configuration of the orbiter and the moon during this third flyby meant that an occultation would take place near the south pole, allowing UVIS to collect data from the southern region for the first time. As the plume moved in front of the light emitted by the background star, UVIS registered dips within the spectra and finally detected water above the south pole.

While the close-up images of the southern area did not reveal the vapor jets streaming from the ridges due to their faint presence, they did provide compelling evidence supporting the remarkable assertion of a plume emanating from Enceladus'southern pole. Now, all that was needed with an image of the plume.

Following the remarkable discovery of the geological activity on Enceladus, the Cassini mission was reconfigured to allow further observation of the moon. And so, the definitive visual confirmation of the plume's existence came just a few months later during a non-flyby encounter on November 28, 2005. In preparation for this new observation campaign, ISS was set with exposures designed to reveal faint atmospheric features and configured with a high phase angle of 148°, a viewing geometry in which small particles became much easier to see. Moreover, various exposure settings and spacecraft rotations were implemented between each set of images to remove any possibility of image artifacts.

The images sent back were simply exceptional. These revealed towering jets that were ejecting vast quantities of particles into space, forming an immense plume which was much bigger than expected, taking everyone by surprise. That a plume of such magnitude and output was emitting from this tiny moon was beyond anyone's wildest dreams. Analysis by the team indicated that the vents were propelling particles to heights of up to 500 km and speeds reaching 2189 km/h, strongly indicating liquid water as the plume's source, given the difficulty of achieving such velocities without liquids.

And so, remarkably, in 2005, Enceladus joined a small group of planetary objects in our Solar System known to harbor active volcanism: Earth, Io, and Neptune's moon Triton. Scientists from around the world started to reconsider their understanding of the Saturnian system with this new discovery.

For instance, the plume emanating from Enceladus provided a solution to a puzzling observation noted earlier by the Cassini spacecraft during its approach to Saturn. Oxygen atoms had been detected across the Saturnian system, the origin of which remained a mystery. However, upon seeing the plume, astronomers understood that the water molecules spewing out from the plume was the source of the oxygen, as radiation would eventually break down the water molecules into their elemental constituents. Furthermore, the discovery of the plume definitively confirmed that Enceladus was indeed the originator of the E ring, composed primarily of icy particles.

It was also found that Enceladus was by far the most significant source of dust, neutral gas, and plasma within the Saturnian system, feeding not only the E ring but also a neutral torus orbiting Saturn. What fiery Io was doing in the Jovian system, Enceladus seemed to be doing something similar in the Saturnian system.

What could account for such activity on this tiny moon? After exploring numerous models, scientists suggested that the vents stemmed from pressurized, heated liquid water situated near the surface.

Flying Through the Plume

The revelation of the vast plume and the enigmatic tiger stripes at the south pole ignited immense excitement within the scientific community and proved to be a turning point for the Cassini mission. Enceladus shot straight to the top of the key targets to investigate in the Saturnian system, and Cassini's trajectory was revised accordingly. While the next—and last—planned flyby (E3) would only occur at the end of Cassini's prime mission 3 years later, on March 12, 2008, it allowed the orbiter to pass at a distance of only 47.9 km through the south region, seven times closer than was originally planned.

At its closest approach, Cassini would be traveling at such high speeds (14.4 km/s) and from such a narrow vantage point that taking shots of the surface during the closest approach would yield little value. Instead, the team had a better idea; they would send Cassini straight through the plume and 'taste' it with its instruments. The objective was straightforward yet ambitious: to analyze the particles emanating from Enceladus, discerning their density, size, composition, and velocity. Meanwhile, the cameras would capture high-resolution images of the surface whenever feasible.

This was a significant change to the original mission. The Cassini orbiter had never been intended for such a feat, and this new trajectory demanded a comprehensive overhaul of the software and a reassessment of the scientific payload's capabilities as these were never designed to analyse samples from the plume. The years leading up to the 2008 flyby were spent in a flurry of preparations for this event.

As this new trajectory was being worked out, an extended mission named Equinox was also being proposed. Consisting of seven new flybys of Enceladus between 2008 and 2010, Equinox had a strong focus on the icy moon and was quickly approved. An additional extension, named Solstice, would ensure the success of the Cassini mission until 2017. The initial flyby within the Equinox extended mission, designated as E4, was scheduled for August 11, 2008, at 54 km from the surface. This would be followed shortly by a second flyby (E5) on October 9, which would bring Cassini to its closest approach ever in the entire 13-year mission, at a mere 25 km from the surface! The final flyby of 2008 (E6), planned for October 31, would position the orbiter 200 km from the surface.

These three flybys shared a similar trajectory: Cassini would first approach Enceladus over the northern hemisphere, then reach its lowest altitude near the equator before passing through the plume over the southern hemisphere. This trajectory allowed different instruments to be activated at different times during the flybys, allowing for a full range of measurements across the moon as the instruments couldn't be activated simultaneously. These repeated flybys could unveil the moon's dynamism by tracking changes observed on the surface over time, particularly at the fissures near the south pole. Navigating through the plume was still deemed risky, so the trajectories were meticulously planned to avoid passing through its most concentrated areas, although it was anticipated that the plume's presence might exert some influence on Cassini, subtly altering its course.

Everyone agreed; 2008 was going to be a big year. In preparation for these four unprecedented flybys, the observation of a new stellar occultation of the plume was made with UVIS in October 2007. This revealed the plume to be composed of four distinct jets, which continued to be tightly concentrated even at an altitude of 15 km, implying that the velocity of the water particles gushing out at the base of the vent had to be at least 2100 km/h.

The team had identified two distinct types of particles being emitted from these jets. One consisted solely of water ice, while the other comprised a blend of ice and non-ice elements. The prevailing belief was that these non-ice elements originated from the moon's interior, possibly from the yet-to-be-confirmed warm subsurface ocean, although not all members of the scientific community concurred with this hypothesis. An alternate theory proposed that a mixture of gas and ice could be trapped beneath the icy crust and, upon exposure to near-vacuum conditions, could erupt into large plumes, eliminating the necessity for warm liquid water altogether. The data gathered by Cassini during the 2008 flybys would enable the scientists to determine the source of the jets.

And so, on March 12, 2008, Cassini opened a new chapter in space exploration by plunging straight through the plume and did so again successfully, flying deeper through the jets during E4, E5, and E6. What it found during these successive flybys provided much insight. The ion and neutral mass spectrometer (INMS) measured the plume's composition: 90% water, 5% carbon dioxide (which suggests that the underground water is carbonated), 1% ammonia (NH_3), 1% methane (CH_4), traces of argon-40 (most likely resulting from the decay of potassium-40 usually found in rocky material), less than 1% of hydrogen sulfide (H_2S), methanol (CH_3OH), and formaldehyde (H_2CO). Other organic compounds might have been detected such as propane, benzene, or hydrogen cyanide, however, limitations in INMS' resolution proved unable to confirm this interpretation at the time. The presence of volatiles such as ammonia and methane are important though, as these could allow water to stay liquid below the freezing point. In addition to this long list of non-ice components, Cassini also detected salts, such as sodium chloride, sodium carbonate, sodium bicarbonate, and potassium chloride, all at concentrations of 1%, equivalent to a tenth of the salinity of Earth's water.

Given that sodium had already been detected in the E-ring particles, finding salt within the plume itself confirmed once more that its source was indeed liquid saline water. Pressurized bodies of saltwater had to be present in the icy crust or underneath it.[2]

In addition to the sampling of the plume, high-resolution images were taken once again at precise regions of the south pole, and as expected, these showed signs of change over time, especially around the seemingly very active tiger stripes. Like mid-ocean ridges on our planet's seafloor, where new material displaces the older crust, the tiger stripes hinted at a similar process and therefore, the presence of significant heat under the icy crust. The detailed images also depicted the process of

---

[2] Despite the subsurface pressures on Enceladus being significantly lower than those on Earth, the moon's small size and relatively low gravity mean that even minimal pressure can propel material far into space.

older vents sealing shut while new ones emerged, a phenomenon seemingly syn-
chronized with the observed fluctuations in the plume. Scientists started to discern
a seasonal pattern in the plume's activity, presumably influenced by Enceladus'orbit
around Saturn.

By analyzing the flyby images, scientists attempted to deduce the age of various
regions by examining their surface features. The northern area, characterized by
heavy cratering, was estimated to be around 4.2 billion years old, aligning with our
understanding of Solar System bombardment phases. In contrast, the equatorial
region displayed a wide range of ages, with some areas as young as 170 million
years and others as old as 3.7 billion years. The south polar region, exhibiting active
surface features, was speculated to be approximately 500,000 years old or even
younger. However, these estimations come with a caveat, as dating objects in
dynamic environments like Saturn's system poses significant challenges for
scientists.

Was there a possibility of a subsurface ocean of liquid water within this tiny
moon? While there was still no conclusive evidence that such a body of water
existed at the time, a growing number of scientists were optimistic that further proof
of its existence would be found.

To better answer this question, the spacecraft, which had already completed 119
orbits around Saturn, was placed on an equatorial orbit allowing for closer and more
frequent flybys of Enceladus. As a result, an additional 13 flybys would be accomplished from 2009 until the end of 2012, by which time Cassini would return to a
highly inclined orbit, preventing any close-up observations of the moon for the following 3 years.

Thus, a year later, Cassini performed a new flyby (E7) on November 2, 2009. Its
nearest encounter brought it within 103 km of the surface, yet this time, it would
venture much deeper into the heart of the plume, as engineers had grown more
assured of Cassini's capability to navigate the jets securely. Nevertheless, because
of the risk associated with this route, engineers decided to use the spacecraft's
thrusters to keep it stable throughout the entire flyby as they carefully monitored its
behavior as it plunged through the plume. If all went well, later flybys could be
completed without using thrusters, thereby preserving precious onboard fuel.
Scientifically, the flyby would provide further insights into the plume's composition, allow the orbiter to map the heat signatures from the surface, and provide
additional high-resolution images of the tiger stripes in the hope of detecting
changes that occurred since the previous flyby (E6).

During E7, Cassini sped through the plume in less than a minute, hurtling at the
neck-breaking speed of 28.8 km/h. The scientific measurements taken during the
flyby helped determine that the plume's density was less than half what had been
predicted. More importantly, though, the plasma spectrometer which was initially
designed to study Saturn's magnetosphere measured the flow velocity of the ions
and electrons within the plume and found negatively charged water molecules. This
supported the idea of a subsurface body of liquid water, as these molecules can be
found when liquid water experiences friction (i.e., a waterfall or crashing waves).
Negatively charged hydrocarbons were also detected in the atmosphere, most likely

the result of Saturn's magnetic field and the Sun's ultraviolet rays interacting with Enceladus' tenuous exosphere. Organic compounds were again detected in the plume, while close images of the south pole helped researchers identify areas that had experienced recent change.

The final flyby of 2009 (E8) occurred on November 2, and was a much more distant flyby than the previous five at 1600 km from the moon's surface. Baghdad Sulcus was the primary focus of the flyby as a detailed thermal map of the fissure revealed heat output throughout the length of the fracture. Temperatures of 180 K (−93 °C) were detected; possibly warm enough for liquid water mixed with ammonia to be present right below the surface. Also, further high-resolution images of the southern hemisphere were taken as well as a full mosaic of the south pole, where 30 individual jets of different sizes can be seen; see Fig. 8.3.

By then, mission engineers had gained enough knowledge and confidence to fly the spacecraft through the plume without the need to activate the thrusters. Such a configuration provided multiple advantages. When the thrusters were on, the engineers couldn't tell if the spacecraft's motion was coming from the thrusters or Enceladus' gravity. With the thrusters off, they could measure the moon's gravitational pull with high precision by accurately tracking the velocity of the spacecraft. The data collected during such a flyby would enable scientists to determine the mass variation under the moon's surface; sometimes called the gravity data.

Fig. 8.3 Enceladus' southern vents and plume captured by Cassini during its close flyby on November 2, 2009. Multiple jets can be seen venting off from the tiger stripes at the south pole. From left to right, they trace out Alexandria, Cairo, Baghdad, and, at the extreme right edge, Damascus sulci. Note that this image has been rotated 180° from its original orientation. (Image courtesy of NASA/ESA)

And so, during the E9 flyby which occurred on April 28, 2010, Cassini flew through the plume once again, but without thrusters. While E9 allowed Cassini to acquire gravity data, future plume flybys planned for 2012 would be required to validate it.

The year 2010 also saw additional close flybys, including an important one on August 13 (E11). During this flyby, new high-resolution images of Enceladus'surface lit by the sunlight reflecting off Saturn were taken, while new infrared data generated the highest resolution heat maps of the tiger stripes yet. This unprecedented set of data revealed warm, complex fractures branching out of Alexandria Sulcus as well as Cairo Sulcus, indicating that significant activity persisted at the end of the stripes.

Furthermore, temperatures of 190 K (−83 °C) were measured within the Damascus Sulcus, which was hotter by 20 K than during the previous measurements made of the same area in 2008. Scientists weren't entirely sure if this was because the fissure was more active or if the 2008 scan, less precise, averaged out the temperatures over the area. Nonetheless, the detection of such high temperatures hinted at substantial geological activity beneath the surface.

Moreover, new high-resolution maps of Damascus Sulcus revealed intricate details as small as 800 m, representing the highest resolution ever achieved of the fissures. These maps also offered the first-ever glimpse of recently ejected material cooling off next to the central trench. Curved striations along the fissures as well as significant temperature variation on its entire length were observed, adding to its complexity. The dynamism within Damascus Sulcus placed it as the most active of all tiger stripes.

Measurements of plume depositions on the surface indicated that even in distant areas from the vents, layers of fine particles were accumulating at a rate of 1 cm per million years. In regions nearer to the tiger stripes, the landscape bore resemblance to snow-covered terrain on Earth, with thick accumulations of plume fallout blanketing the surface.

With 2010 coming to a close and winter setting in the Saturnian system, enveloping Enceladus'south pole in darkness for several years ahead, observations of the plume and the tiger stripes would be more difficult.

The Evidence for a Subsurface Ocean

In 2011, new studies of the data acquired during the 2008 and 2009 flybys confirmed the presence of large grains of ice containing substantial amounts of salt, reaching concentrations of up to 2%. This contradicted previous findings that had proposed smaller ice particles with lower salt levels. Given the known solubility of salt in water, it was deduced that these larger icy grains likely stemmed from a reservoir of saline liquid water, lending further support to the existence of a subsurface ocean.

Additional flybys in 2010 and 2011 would provide new measurements of the moon's gravitational pull as well as new surface images and further plume sampling. A total of 90 jets varying in size were identified over the south pole, all sprouting from the tiger stripes.

Starting in 2012, Cassini was returned to a highly elliptical orbit, precluding any flybys for the subsequent 3 years. In fact, only three close encounters with the moon were to take place, all in 2015, before the conclusion of the Cassini mission in 2017. In the meantime, scientists were busy analyzing the finer details of the data collected by the numerous encounters with the plume.

For example, a study conducted in 2013 established a correlation between the fluctuations in the plume's activity and Enceladus'position in its orbit around Saturn. Most planetary orbits are not perfect circles, but instead, exhibit eccentricity; their orbits are elliptical rather than circular. The study demonstrated that when Enceladus was at its farthest distance from Saturn, the intensity of the plume increased threefold compared to when it was closest to the planet. This observation leaves little doubt that as the moon completes its orbit around Saturn, the gravitational variances resulting from its elliptical path are sufficient to compress and stretch the icy crust. These changes cause the openings of the south pole fissures to vary.

The further the moon distances itself from its primary planet, the wider the openings of the fissures become. As the moon gets closer though, the surface gets tighter, limiting the openings and the amount of material that can be vented out from the jets. This new insight would allow the Cassini team to better plan their future observations of the moon. Interestingly, this pattern might also occur on Jupiter's moon Europa and could help explain why observations of Europa's hypothetical plumes by the Hubble Space Telescope seem to be intermittent throughout the years.

In 2014, a study by Luciano Iess and his colleagues from the University of Rome provided the long-awaited evidence of a subsurface reservoir of liquid water, a discovery eagerly anticipated by scientists for years. This breakthrough stemmed from gravimetric measurements conducted during three flybys over both the south and north poles of Enceladus between 2010 and 2012. Initially, it was assumed that the southern region, with its significant depression as indicated by topographic data, would exhibit a weaker gravitational force compared to the north pole. However, the findings revealed the opposite: the south pole region exhibited a much stronger gravitational pull than anticipated, indicating a greater mass directly beneath it. The study revealed that only liquid water, which is denser than ice, would fit the gravitational observations, thus confirming the presence of a large body of liquid water under the south pole. Estimated to be the size of Lake Superior in North America, the subsurface sea was thought to cover the entire southern region and be under 30–40 km of icy crust. The models estimated the sea to be 8–10 km deeper, and crucially, that it was resting on the moon's rocky mantle. Water and rock interactions were possible.

After years of tantalizing clues and speculations, scientists finally managed to conclusively prove that Enceladus was indeed an ocean world, albeit initially perceived as a regional sea rather than a global ocean.

This news captivated the public and scientists alike since much of what life needs to thrive was present on Enceladus: liquid water, a heat source, organic compounds, and the presence of minerals through the water/rock interaction. One unknown parameter was the age of the moon's south polar sea. Contrary to Europa's subsurface ocean, which is estimated to have existed for billions of years, determining the age of Enceladus' subsurface sea was difficult at this stage. However, some clues hinted that it might be young.

For example, the rate of argon-40 spewed out into space by the vents can be measured and compared with the theoretical estimate of the amount of argon-40 generated from the moon's rocky mantle. Calculations have suggested that at the current venting rate, the reservoir of argon-40 would have only lasted ten to a hundred million years, giving the impression that the plumes might be periodical or that this is a one-off event that we are lucky to observe.

The planetary scientists were still grappling with further mysteries. Given the heat measured within the southern region, estimates had Enceladus' total energy output at 16 gigawatts, which is far more than what the moon is estimated to receive through radioactive decay and tidal heating. Additionally, the unusually high activity at the south pole compared to the north pole puzzled researchers. Tidal heating models suggested a consistent heat output on both poles, yet observations revealed otherwise. Clearly, there was much more to uncover about this intriguing moon.

In 2015, following a 3-year hiatus, Cassini's orbit around Saturn permitted it to resume its close approaches of the icy moon. However, only three planned flybys of Enceladus would be conducted before the conclusion of the Cassini mission.

During the second of these close flybys (E21), which occurred on October 28, the probe flew through the plume at a distance of 49 km, the lowest it had ever approached the south pole. E21 marked the final occasion Cassini would journey into Enceladus'plume itself, as the subsequent flyby, E22, followed a different trajectory.

The scientists were looking for two things during E21. Firstly, by flying so low within the plume, they would expect Cassini's instruments to sample heavier organic compounds, which due to their weight, did not rise to the altitudes attained by the previous flybys. Secondly, it was hoped that molecular hydrogen might be detected in the plume, as this would be evidence that hydrothermal activity was ongoing on the seafloor. Also, such studies would allow scientists to extrapolate how much heat was being produced by the hypothetical deep-sea vents, and whether these could be conducive to life. The results of the hydrogen investigation would be confirmed a few years later.

The E21 flyby took place as expected, and while the data collected during the approach was being investigated, a new research revealed that nanometer-sized particles of silica were found floating freely within Saturn's giant magnetosphere and were thought to originate from Enceladus' plume. Silica, also known as silicon dioxide (SiO_2), is a major constituent of sand and mostly found in rocks, which strengthens once more the case for water/rock interactions. Interestingly, hot liquid environments are required for silica particles to be formed, with some studies suggesting water temperatures at the base of the subsurface ocean could reach above 50 °C. Since such high temperatures can only be attained through hydrothermal activity, scientists concluded that the silica particles might have been formed by hot hydrothermal vents at the base of the seafloor. These silica particles were then transported to the surface and vented out into space with the water-ice.

In the same year, another significant research paper emerged. Led by Peter Thomas, a former member of the Cassini mission's imaging team, the study delved into 7 years of Cassini data on Enceladus. Its findings suggested that the subsurface body of water wasn't limited to the south pole; instead, it enveloped the entire moon, creating a global layer of liquid water. The study analyzed hundreds of images of the moon's surface taken over the years and mapped the positions of the craters observed on the surface with extreme precision. This allowed scientists to accurately measure the way the surface was moving while the moon rotated on its axis. Peter Thomas and his colleagues found that the surface was rotating at a different rate compared to the moon's interior, indicating the presence of a global layer of liquid water separating the rocky mantle from the icy crust. In effect, Enceladus' icy crust is decoupled from the moon's interior and rotates independently as it orbits Saturn. By carefully tracking the crust' rotation, referred to as libration, extrapolations can be made of the moon's interior, with early estimates suggesting that the ocean was 30 km deep on average (Fig. 8.4).

Two challenging questions remained unanswered. Firstly, what was the age of this subsurface ocean? Secondly, and equally crucial, how did it originate, given that the energy output from tidal heating and radiogenic heating alone doesn't appear sufficient to form a global ocean? As scientists mulled over these questions, additional data continued to accumulate.

A few days before Christmas, on December 19, 2015, Cassini performed its final encounter with Enceladus (E22) flying at a distance of 5000 km from the surface, placing it at an ideal location to map the southern hemisphere and measure once

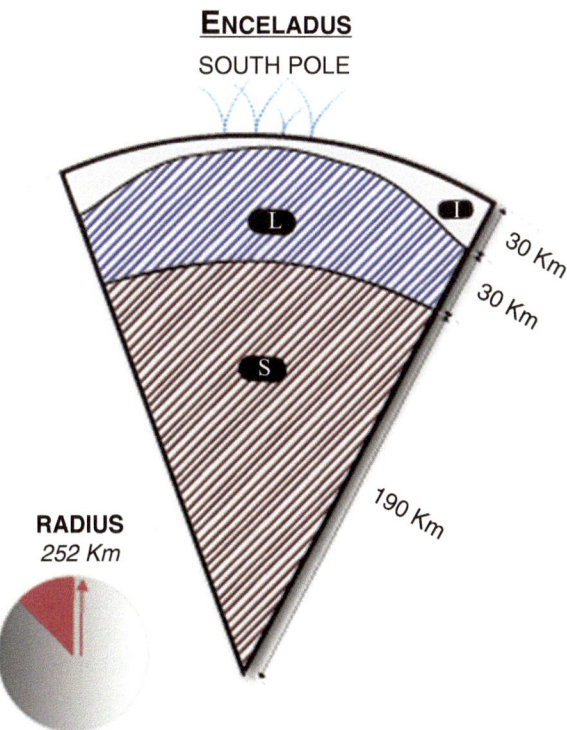

Fig. 8.4 Diagram showing the interior of Enceladus. The thickness of the icy crust averages 30 km but gets very thin, at 1 km (or even less) at the south pole. *I* Ice, *L* Liquid, *S* Silicate mantle. Diagram is not to scale.

again the heat flow at the south pole. The Cassini team, though saddened by this final encounter with the remarkable moon, found solace in the realization that the data acquired by their spacecraft would keep scientists busy for decades to come. And so, Cassini said its adieu to Enceladus with a faultless last flyby.

A year later, in 2016, a study utilized data from prior years to revise the estimates of Enceladus'ice shell thickness, suggesting it to be approximately 35 km thick at the equatorial regions and 5 km thick or less at the south pole.

The Discovery of Hydrogen

The most exciting discovery was yet to come through. April 2017 saw NASA beat the drums by organizing a press event; hydrogen—gasp—had been detected in the data from the E21 flyby a few years earlier. In fact, Cassini had already detected hydrogen in the past whenever it had analyzed the plume. However, scientists had been unable to determine if the detected hydrogen had originated from the moon itself or the scientific instrument. Indeed, the spacecraft's ion and neutral mass spectrometer (INMS), which measured the hydrogen, wasn't originally designed to perform under such conditions as its original purpose was to sample the upper atmosphere of Saturn's moon Titan.

Whenever particles from Enceladus'plume would be sampled, there was always a possibility that they would interact with the titanium walls inside the instrument and produce hydrogen. To counter this uncertainty, the team behind INMS worked hard prior to the E21 flyby to reconfigure the instrument and prevent incoming molecules from coming into contact with the titanium walls. With a newly configured INMS, the E21 flyby collected valuable data and hydrogen was finally confirmed. This took most scientists by surprise though, as a large amount of hydrogen was detected (1%). This suggests that a continuous production of hydrogen molecules (H_2) must be occurring. The presence of hydrothermal vents located at the bottom of the subsurface ocean would likely be the culprit, confirming the interpretation made earlier based on the presence of silica particles within the plume.

If there are vents on Enceladus'ocean floor, they are most likely alkaline hydrothermal systems similar to the 'low temperature hydrothermal systems' present on Earth's ocean floor. These systems are distinctly different from the famous "black smokers" found in Earth's mid-ocean ridges, as they are heated not by volcanic energy, but through exothermic reactions between seawater and rocks. This reaction is called serpentinization and makes the vents less hot (40–75 °C) and alkaline (pH 9.0–9.8). Carbonate chimneys dominate these low temperature hydrothermal vents, providing vast feeding grounds for dense microbial communities. The heat released by the serpentinization contributes to the circulation of water through the rocks, feeling further reactions. Furthermore, the rocks subjected to serpentinization undergo a chemical alteration during this process, reducing their density and increasing their volume. Ultimately, this process leads to their separation from the parent rock. This serpentinization can widen existing fractures and generate extensive areas of interaction between water and rock surfaces.

The discovery of hydrogen within Enceladus' plume is also important because hydrogen is one of the essential building blocks for life as it forms the basis of organic compounds on Earth and is a food source for life forms. Indeed, hydrogen is used by microorganisms (methanogens) to produce energy by converting hydrogen and carbon dioxide into methane. As a prominent NASA scientist exclaimed, the discovery of hydrogen is like finding a candy store for microbes. We will explore the implications for the moon's habitability a bit later.

To address the energy puzzle of why there is more energy within Enceladus than anticipated based on the known characteristics of the moon, two intriguing theories emerged in 2017. These ideas not only offer potential explanations for this question but also begin to elucidate the presence of a subsurface ocean.

The first theory posited a scenario involving a violent collision of the icy moon with a large rock. In a study published in 2017 by Angela Stickle at Johns Hopkins University in Maryland and John Spencer at the Southwest Research Institute in Colorado, it was demonstrated that a significant impact around a hundred million years ago could account for the observed heat output and massive fissures at the south pole. Their calculations indicated that a sufficiently powerful strike capable of penetrating the presumed 20 km thick icy crust at the time would have deposited substantial energy within the impact site, leaving enduring scars on the ice shell.

The impact itself didn't necessarily need to occur directly at the south pole. Such a collision would have caused a gravitational anomaly at the impact site, prompting the moon to reorient itself to shift the crater toward one of its poles, a phenomenon similar to what occurred with Pluto, which will be discussed in Chap. 9. In this instance, the impact site just happened to be closer to the south pole than the north. Intriguingly, some studies have proposed that Saturn's rings, primarily composed of icy particles, could also be approximately a hundred million years old, formed from the breakup of a small icy moon. These two hypothetical events may be interconnected.

Another concept suggested for the significant energy output of the moon emerged in 2017 from research conducted by Gaël Choblet and Gabriel Tobie, both affiliated with the University of Nantes in France. Their studies proposed that if Enceladus' core consisted of loose, easily deformable porous rock, cold water from the ocean could infiltrate it, gradually warming up due to tidal friction as water interacts with the surrounding rocks, and then ascend back up through convection. This process of heat transfer from the core to the ocean has been demonstrated to generate sufficient energy to sustain a plume for billions of years, if not longer.

Further analysis in 2019 was done of the spectral characteristics of ice grains in the plume. The research indicated the probable presence of nitrogen-bearing and oxygen-bearing amines, carrying significant implications for the potential availability of amino acids within the internal ocean.

Another study on the plume's ice grains came in 2023 as scientists examined archived data collected by the Cosmic Dust Analyzer (CDA) while Cassini was passing through the E ring. By using more recent modeling of mineral solubilities and laboratory analogue experiments, the scientists showed that CDA detected phosphate salts (Na_3PO_4 and Na_2HPO_4); in other words, phosphorus. This discovery is significant because phosphorus plays a vital role in biological processes. In fact, phosphorus completes the ensemble of essential elements found in Enceladus'subsurface, providing the six key elements for life: CHNOPS (see Chap. 3). This finding also indicates that the ocean is rich in phosphates, an unavoidable result of the interaction between alkaline and carbonate-rich water and rock, confirming once more the continued geological activity at the bottom of the ocean.

Further examination of Cassini's data continues to unveil new discoveries. In 2023, additional precursors of life were identified within the ion and neutral mass spectrometer (INMS) dataset. While numerous compounds had been detected, a problem persisted. Due to the inherent limitations of INMS, and more precisely, its low mass resolution, the identification of additional compounds detected by the instrument proved difficult as a large number of plausible models were compatible with the data. Recently, to resolve these ambiguities and effectively deal with the complexity of plume models, scientists have refined their statistical modelling and published a paper confirming the identification of new compounds in the sampled plumes. Thanks to these new models, the new compounds identified with very strong certainty include hydrogen cyanide (HCN), acetylene, propane, and alcohols. While poisonous to most creatures on Earth, HCN could have played a key role in chemical reactions that created the ingredients that set the stage for the advent of life. Indeed, some biologists believe that it may have acted as a precursor to nucleic acids and amino acids. In addition, detecting acetylene and propane in the plume further confirms ongoing catalytic reactions driven by minerals within the ocean.

Enceladus' Habitability

On September 15, 2017, the orbiter's fate was sealed. Running out of fuel for adjusting its course, it was purposefully plunged towards Saturn to avoid any risk of contaminating Enceladus or Titan if left unchecked. And so, high up in the gas giant's upper atmosphere, one of the most successful spacecraft ever to be sent into space disintegrated in a fiery dive.

For more than a decade, the Cassini-Huygens mission characterized Enceladus and its plume in great detail. Multiple lines of evidence have revealed the presence of a warm and salty subsurface ocean circumventing the entire globe under a 35 km thick icy crust, which is much thinner at the south pole, as little as 1 km perhaps. This ocean is on average 30 km deep and in direct contact with a rocky mantle. It contains organic compounds of various lengths such as among others: methane (CH_4), propane (C_3H_8), methanol (CH_3OH), formaldehyde (H_2CO) as well as salts, carbon dioxide (CO_2), ammonia (NH_3) and methane (CH_4) both acting as an antifreeze, hydrogen cyanide (HCN), hydrogen sulfide (H_2S), silica (SiO_2), argon (isotope argon-40), hydrogen (H_2), and more. All six essential elements vital for life as we know it—carbon, hydrogen, nitrogen, phosphorus, sulfur, and oxygen, referred to as CHNOPS—have been found in the plume and the E ring which it feeds.

This energy drives the eruption of over a 100 jets, expelling approximately 200 kg/s of both ice and non-ice particles through extensive fissures called tiger

stripes. Many of these icy particles reach escape velocity, contributing to the formation of the expansive E ring encircling Saturn, while the heavier particles fall back
onto the moon's surface.

The existence of low-temperature hydrothermal vents, indicated by the presence
of hydrogen in the plume, suggests that Enceladus'ocean may be one of the most
hospitable extraterrestrial environments in our Solar System. Astrobiologists envision the possibility of microbial ecosystems, potentially based on methanogens,
thriving near the base of these ocean floor vents. Methanogens utilize carbon dioxide and hydrogen expelled by the hydrothermal systems to produce energy, generating methane and water as by-products. The reaction is written like this:

$$CO_2\left(aq\right) + 4H_2\ \left(aq\right) \rightleftharpoons CH_4\left(aq\right) + 2H_2 0\ \left(l\right)$$

In Chap. 6, we saw how this reaction might not occur on Europa given the
extreme pressures encountered at the bottom of the Europan subsurface ocean
whose depths reach to 100–150 km. On the other hand, the conditions within
Enceladus' smaller ocean are still favorable for this reaction to take place, ensuring
an ample supply of food for the microorganisms.

However, the sheer abundance of hydrogen found in the plume raises eyebrows
among the scientific community. Jonathan Lunine, a professor at Cornell, humor-
ously dubbed it the "Neglected Pizzeria," evoking the image of fresh pizzas stacked
up high in a pizzeria. This analogy prompts a pertinent question: Why is no one
consuming the hydrogen? Indeed, one could argue that an excess of hydrogen suggests that life may not exist in Enceladus'subsurface ocean, as any potential life
forms would likely have used it by now.

While methanogenesis represents a widespread microbial metabolic process utilizing chemosynthesis, on Earth we also observe ecosystems where abundant hydrogen remains largely untouched by life. This can occur due to physical constraints
that restrict its consumption, such as low levels of phosphorus essential for basic life
functions, or the presence of alternative food sources to which microorganisms have
adapted. As a result, some astrobiologists do not view the surplus of hydrogen as a
compelling argument against the potential for life on Enceladus.

There is however one aspect of Enceladus' subsurface ocean where great uncertainty remains: its age. The complexity inherent in modeling the formation of the
Saturnian system with its rings and multiple moons makes it hard to determine
Enceladus' age. The tiny moon might have formed with Saturn 4.5 billion years ago
or much later, around two billion years ago. Relying solely on the count of surface
craters for age estimation introduces the risk of misinterpretation, as the rate of
impacts may have fluctuated over time in ways that remain poorly understood. For
instance, recent studies suggest that Saturn's rings, previously believed to be billions of years old, may actually be only around a hundred million years old.

Furthermore, some scientists suspect that the subsurface ocean might be far
much younger than the moon itself. Some place it at only 500 million years old,
while others put forward even shorter timescales, implying a periodicity in the formation of the subsurface ocean. Currently, the age of Enceladus remains uncertain,

with estimates ranging from potentially hundreds of millions to billions of years old. The reality is that there are still many unknown factors, and any conclusions regarding the ocean's age have yet to be reached.

Nevertheless, what do these younger estimates imply for the possibility of life in Enceladus' subsurface ocean? Although we have seen in Chap. 3 that life did require hundreds of millions years to start off on Earth, some scientists view the young age of Enceladus' ocean as a good thing. If the moon and its subsurface ocean was old, estimates have suggested that the serpentinization process that we observe today should have run out, and with it, the nutrients required for life. Indeed, given the moon's small size, there might not be enough material to sustain life for significant periods of time. In the case of Enceladus, a younger ocean filled up with nutrients and energy might be a better scenario than an old one where life is starved of nutrients.

Should the ocean on Enceladus indeed be young and conducive to life, any organisms that may have arisen from it would likely be simple life forms, such as single-cell organisms. The evolution of complex cells or multicellular organisms seem to require billions of years as was the case on Earth.

To detect such life, ambitious missions have been proposed to sample the plume material and analyze them using state-of-the-art laboratories onboard a spacecraft or even, to return plume samples directly back to Earth for extensive analysis in the best laboratories in the world. Ideally, a sample return mission would provide the best scientific return. Chapter 12 will explore such missions in detail.

Contemplating the prospect of a sample return mission from Enceladus raises several significant implications. If such a scientific endeavor were to reveal life in the subsurface ocean similar to life on Earth, it would strongly support the panspermia hypothesis. Conversely, if the life discovered on Enceladus were markedly different from life on Earth, it would suggest that hydrothermal vents on seafloors might be the natural environments for life to emerge on planetary bodies. Alternatively, if no life were found within Enceladus'subsurface ocean, it might indicate that additional conditions, such as those provided by Earth's atmosphere, are necessary for life to arise. Even if we discover that the third scenario is the right one, it will be truly fascinating. Learning about the boundaries imposed on life will provide much insight into how life appeared on our planet and, ultimately, where we come from. Therefore, searching for life within the icy moon of Enceladus is also, in a way, searching for our own origins.

As we bid farewell to this captivating and promising moon, we conclude the second leg of our journey, which has taken us to the five confirmed ocean worlds in our Solar System: Ganymede, Callisto, Europa, Titan, and Enceladus. Now, we turn our attention to the third part of our exploration, where we delve into planetary objects that present intriguing hints of subsurface oceans or bodies of liquid water yet require further evidence to confirm. The upcoming section will shine a spotlight on two dwarf planets and three moons, each offering exciting prospects for expanding our understanding of habitable environments within our Solar System. Without delay, let us embark on our expedition to Ceres, Dione, Ariel, Triton, and Pluto.

Part III
Possible New Ocean Worlds

Somewhere, something incredible is waiting to be known. (Carl Sagan)

In this part, we review the planetary objects that might harbor subsurface oceans or small bodies of water but for which we still await confirmation. We will cover a wide range of Solar System objects, from moons to dwarf planets, from Kuiper Belt objects to far away Scattered Disk objects. To highlight the variety of objects presented in this part, we will start in Chap. 9 by reviewing three very different objects: the dwarf planet Ceres, Saturn's moon Dione, and Uranus' moon Ariel. In chapter ten, we will continue with Neptune's moon Triton and dwarf planet Pluto, which share numerous similarities. In Chap. 11, we will explore objects that in theory could have harbored a subsurface ocean in their past or may still do so today.

Chapter 9
Ceres, Dione, and Ariel

Ceres is a dwarf planet located in the asteroid belt while Dione and Ariel are moons of Saturn and Uranus respectively. Despite residing in vastly different corners of our solar system with share some intriguing characteristics. All three are icy worlds, composed primarily of water ice and rock, suggesting that they formed in the colder, outer regions of the early solar system. These three bodies also have intriguing surface features making them strong candidates for hosting subsurface bodies of liquid water.

Ceres

Discovery and Observations

In 2006, the International Astronomical Union (IAU), the organization that represents the majority of professional astronomers around the world, held a conference in the picturesque capital city of Prague in the Czech Republic. Although this event was famous, or ill-famed depending on your point of view, for demoting Pluto to a dwarf planet and in the process making millions of schoolbooks and bedroom posters obsolete overnight, news was also made on the change of status of another Solar System body: Ceres (official known as 1 Ceres). Until then, Ceres held the distinction for being the biggest asteroid in the Asteroid Belt, a ring made up of millions of asteroids orbiting the Sun between Mars and Jupiter. It now became part of a select club of 'Dwarf Planets' alongside four other objects: Pluto, Eris, Makemake, and Haumea.

Ceres was discovered in the early nineteenth century, a period where the enthusiasm for astronomy was flourishing in Europe. French astronomer Charles Messier had published his famous astronomical catalogue a few decades earlier, which was

© The Author(s), under exclusive license to Springer Nature Switzerland AG 2024   171
B. Henin, *Exploring the Ocean Worlds of Our Solar System*, Astronomers' Universe,
https://doi.org/10.1007/978-3-031-62953-2_9

soon followed by Uranus' discovery by William Herschel. With such keen interest in observing the night sky, discoveries were bound to occur.

On the first day of the year 1801, Guiseppe Piazza, an Italian priest, mathematician, and astronomer based in Sicily, was observing the night sky searching for a particular type of star when he spotted a "slow-moving star-like object." He initially thought it was a comet but had some reservations, since these objects were known at the time to be moving fast throughout the sky, which was not the case with this new object. By an ironic twist of fate, the new object the Italian astronomer had found was located exactly where the Titius-Bode law—a now discredited theory published a century earlier—had predicted a planet to exist. Titius and Bode, two German astronomers, were convinced that a distinct pattern existed within the mean distances between the planets and the Sun known as their semi-major axis. They had envisaged, with the Scottish mathematician Colin Maclaurin, that a small planet should be present between the orbits of Mars and Jupiter. When Piazza's discovery of a new object within that precise location was announced, astronomers championing the Titius-Bode law were convinced that he had found the missing planet and duly accepted Cerere Ferdinandea, after the Roman goddess of Agriculture, as the name chosen by the Sicilian astronomer.

Therefore, for half a century, astronomical books and charts showed Ceres as a planet. Nevertheless, it wasn't the only object lying in this particular area, and as observations of the night sky carried on, so did new discoveries. By the 1820s, astronomers counted 11 planets in the Solar System: Mercury, Venus, Earth, Mars, Vesta, Juno, Ceres, Pallas, Jupiter, Saturn, and Uranus. In a prelude to the recent demotion of Pluto, by 1847 astronomers came to realize that Ceres and its five other companions (Vesta, Juno, Pallas, Iris, and Hebe) were part of a new category of Solar System objects that Herschel termed asteroids, meaning 'star-like' in Latin due to their apparently being indistinguishable from regular stars. The Asteroid Belt was introduced, and the Solar System reverted to seven planets once more.

We now know that the Asteroid Belt is comprised of millions of irregular objects made up of rocks, metals, and some ices. Nestled within this belt, the dwarf planet Ceres stands out as being the biggest asteroid, making up almost a third of the total mass of the Belt itself; Vesta makes up 9%, Pallas 7%, and Juno 1%. Through their calculations, astronomers knew that Ceres was big enough, at 473 km radius, for the gravitational pull of its mass to shape it into a ball, yet it was still too small for astronomers to see it as a disc and study it properly. Nevertheless, it was suggested that due to its size, Ceres could have experienced differentiation, the separation of the constituents of a planetary body creating distinct layers within its interior, and a large amount of primordial water-ice might be present as an icy subsurface mantle.

As technology improved, plans to study Ceres started in earnest in the 1970s and 1980s, and characteristics such as albedo and spectrum in visible and near infrared, suggested that Ceres must be similar to carbonaceous chondritic meteorites. This type of meteorite represents primordial matter that has been relatively unaltered by heating throughout its history. The metal in these meteorites are mainly silicates, oxides, or sulfides, with most containing water and minerals, as well as organic compounds. As a broad principle, the outer region of the Asteroid Belt harbors objects like carbonaceous chondrites, which haven't experienced the intense heat prevalent during the early stages of the Solar System. In contrast, those found in the inner part of the belt, such as Vesta, have been subjected to substantial heating and resemble more the silicate rocks found on Earth.

As further observations were made, astronomers improved the estimates of Ceres' mass and radius, leading them to calculate its mean density at 1.98 +/– 0.03 g/cm^3, halfway between that of water (1 g/cm^3) and the average rock (3 g/cm^3). Given its density, researchers concluded that the giant asteroid must be composed of at least a quarter water.

Despite these observations though, many questions remained unanswered such as the composition of Ceres' surface and subsurface layers, the properties of its regolith, or its degree of differentiation. Thankfully, the arrival of powerful new telescopes in the 1990s and 2000s, would bring us closer to answering these questions. Already, in June 1995, the Hubble Space Telescope (HST) took the first direct albedo maps of Ceres in the hope of detecting a possible polar cap. The images taken in ultraviolet light, see Fig. 9.1, revealed details 50 km across that, despite their fuzziness, suggested the existence of a dark spot 240 km in diameter on the surface, most likely a giant crater.

Seven years later, ground-based observations from the powerful Keck telescopes provided sharper images and confirmed the presence of the dark spot detected by HST and a few more darkish areas. More importantly, observations made by the Earth-based telescopes established that Ceres is an oblate object. Since the shape of a spherical object depends, among other things, on its rotation speed and the mass distribution within its interior, scientists used models to infer that Ceres' interior could be differentiated with an icy mantle resting on a rocky core. Some even went

Fig. 9.1 Ceres observed by the Hubble Space Telescope in 1995. It was the first time the asteroid was seen as disc. Fuzzy areas show a darkened area in the center. (Image courtesy of HST/NASA/ Southwest Research Institute)

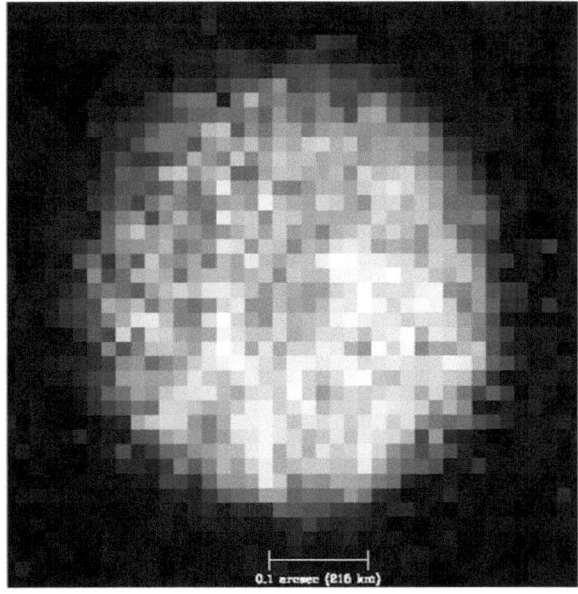

0.1 arcsec (216 km)

as far as to predict a 100 km thick layer of water-ice. Given this, Ceres was thought to be a large, wet protoplanet.

During this time, the HST underwent multiple repairs and upgrades, significantly improving its capabilities (Shuttle Service Mission 2 in 1997, Service Mission 3A in 1999, and Service Mission 3B in 2002). Thanks to these updates, new colored images of Ceres taken by HST in 2003 and 2004 covered a full rotation of the dwarf planet (9 h) and showed multiple recognizable surface features and mysterious bright spots (Fig. 9.2).

The giant asteroid got even more remarkable when a 2005 study by Thomas McCord at the Hawaii Institute of Geophysics and Planetology, University of Hawaii, and Christopher Sotin, at the time at the Laboratory de Planetologie et Geodynamique, University of Nantes in France, calculated that the radiogenic heating within the protoplanet would be sufficient to create and potentially sustain a small subsurface ocean if enough differentiation had occurred and an icy mantle was present.

By then, the scientific community was actively advocating for a spacecraft to explore the dwarf planet and conduct detailed examinations. Fortunately, the Dawn mission, aimed at investigating the asteroids Ceres and Vesta, was chosen in 2001 for NASA's Discovery Program. Despite encountering numerous delays and setbacks, the mission was ultimately launched in 2007.

While Dawn was en route to the Asteroid Belt, a new study published in 2009 by Mikhail Zolotov of Arizona State University proposed that, contrary to having an icy mantle, Ceres might be relatively dry, as its low density could also be explained if it was an undifferentiated body consisting mainly of porous rock increasing in density as we approach the core. The competing theories—dry and undifferentiated

Fig. 9.2 Ceres imaged by the Hubble Space Telescope in 2003 and 2004. (Image courtesy of NASA/ESA/J. Parker [Southwest Research Institute], P. Thomas [Cornell University], L. McFadden [University of Maryland, College Park], and M. Mutchler and Z. Levay [STScI])

versus wet and partially differentiated—would split the scientific community into two camps for many years.

Thankfully, they didn't have to wait too long. Prior to Dawn's arrival at Ceres in 2015, observations made a year earlier by ESA's Herschel Space Observatory (a Hubble equivalent in infrared), hinted that water vapor was escaping from two specific areas linked to mid-latitude regions on the surface. Such events could either be due to comet-like sublimation, where ices are transformed into gases, or to cryovolcanism, where internal heat creates ice geysers similar to what is taking place on Enceladus (see Chap. 8). This discovery fired up the scientific community. When the Dawn spacecraft inserted itself into Ceres' orbit in the spring of that year, a new chapter in the exploration of protoplanets had begun.

The Dawn Revolution

In many ways, Dawn was a pioneering mission. It was the first exploratory space mission to use ion propulsion, which allows spacecraft to enter and leave the orbit of multiple Solar System objects. Before arriving at Ceres, the spacecraft had already orbited Vesta, another noteworthy protoplanet in the Asteroid Belt. Dawn was also the first spacecraft to visit Vesta and Ceres.

The spacecraft had three instruments. Two of its three main scientific instruments were provided by Europeans; the framing camera (FC) was built by the German Space Agency, and the visible and infrared spectrometers (VIR) were built by the Italian Space Agency. The other scientific instrument was the gamma ray and neutron detector (GraND) built by the Los Alamos National Laboratory in the U.S.

Dawn concluded its mission at Ceres in 2018 after nearly 3 years of intensive study, capturing over 95,000 images and comprehensively mapping its surface (Fig. 9.3). Throughout its mission, the orbiter unveiled a captivating landscape characterized by salty brines, icy formations, and remnants of previous activity, including ancient cryovolcanoes. Let's delve into some of the most remarkable findings.

While Dawn is littered with numerous young craters, astronomers were surprised by the lack of large craters on its surface, with none bigger than 280 km in diameter. The scarcity of large craters contradicts expectations for an ancient Solar System object aged 4.5 billion years. This anomaly presents two potential explanations: either ongoing geological activity, such as ice volcanism, has resurfaced the dwarf planet, or underlying layers of ice or low-density material, such as salt, have gradually smoothed surface features over time, obscuring large craters.

What also became immediately apparent in the first pictures taken by Dawn was the presence of very bright spots, called faculae, standing out on an otherwise coal-dark surface. The spacecraft has detected over 130 such spots, most of them within craters, with the brightest lying in Occator Crater, a 90.5-km-wide impact crater formed around 22 million years ago. Recent measurements estimate the age of the faculae in Occator Crater as being only 4 million years old, the blink of an eye in geological time.

Fig. 9.3 Ceres viewed by
the Dawn spacecraft in
2017. This high-resolution
image provides a full view
of the dwarf planet, with a
'faculae' clearly visible as
a bright spot within the
Occator Crater in the
middle of the image.
Recent studies estimate the
age of the cryovolcano at
around a few million years
only. (Image courtesy of
NASA/JPL-Caltech/
UCLA/MPS/DLR/IDA)

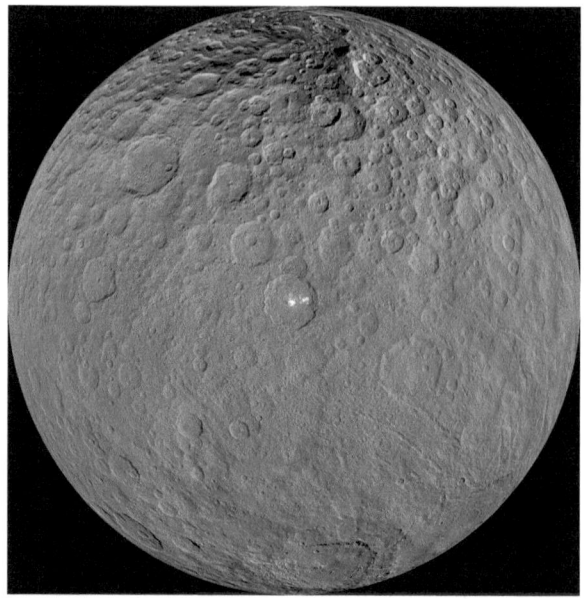

The nature of the bright spots is not icy material as was initially suggested, but instead, large accumulations of sodium carbonate. In fact, Ceres has the largest deposits of sodium carbonate discovered outside of Earth. Intriguingly, on Earth, this mineral is formed in aqueous conditions linked to a hydrothermal environment. Warm liquid water is therefore required to bring these minerals up to the surface, most likely in the form of brine. Although this could be suggestive of a very wet subsurface, it is worth pointing out that minimal amounts of moisture within the surface can also generate brines.

Additional evidence supporting the notion of Ceres possessing a wet interior has been gathered. Ahuna Mons, an impressive yet lonely 4 kilometers high ice volcano which appears to be only a few hundred million years old, seemed to have been formed from ices containing substantial concentrations of salts whose nature is unknown at present. Its origin is believed to stem from hot subsurface brine surging to the surface via fissures in the crust, gradually accumulating over time to construct the massive volcanic dome.

Dawn also confirmed the existence of a transient atmosphere mainly composed of water molecules, supporting the 2014 observations by ESA's Hershel telescope. This atmosphere seems to be present only during intense solar activity, which is thought to warm up and sublimate exposed surface ice within the craters, thus generating a seasonal exosphere. This is similar to what has been observed in comets and provides a direct line of evidence that water-ice is still present on the surface of the dwarf planet.

Additional evidence for the existence of water-ice on Ceres' surface would be published in 2016, as Dawn's gamma ray and neutron detector (GraND) found hydrogen in the uppermost surface, indicative of the presence of water. GraND effectively monitors the neutrons emitted from Ceres' surface as it interacts with galactic cosmic rays (GCRs), which are highly energetic particles from space constantly bombarding the protoplanet's surface. These neutrons do not disperse randomly into space; instead, their distribution follows a pattern determined by the composition of the material struck by the GCRs. By monitoring the flux of GCRs bombarding the surface and analyzing the neutrons produced, researchers can deduce the nature of the materials struck by these cosmic rays. Impressively, GraND has the capability to detect neutrons originating from material located just beneath the surface, up to a depth of about a meter, thereby providing insights into both surface composition and the immediate subsurface.

With such capabilities, Dawn successfully detected subsurface water distributed throughout Ceres, with significant concentrations observed particularly at mid to high latitudes where temperatures are relatively lower. Such a discovery doesn't necessarily imply that a thick icy crust, like the ones found within the icy moons of the outer planets, exists on Ceres. The measurements could also indicate a combination of water-ice and porous, rocky material, potentially forming an undifferentiated crust composed of both rocks and ice. Dawn has also detected a few patches of water-ice lying directly on the surface, although these are located within areas of permanent shadow within crater rims, although the total amount of water-ice on the surface is marginal compared to the subsurface water detected by GRaND. Intriguingly, Dawn also found the presence of NH_3 on the surface in the form of ammonia-rich clays and ammonia salts. Since ammonia condensed within the outer Solar System, far away from where Ceres currently lies, either the dwarf planet was formed in the outer Solar System and then migrated to its current position within the Asteroid Belt or underwent significant bombardment in the past by outer Solar System objects abundant in ammonia. The intriguing idea that Ceres originated in the outskirts of the Solar System is not new as planetary scientists have posited for years that it shares more similarities with distant objects like Pluto than with the asteroids found within the asteroid belt itself. The early formation of the Solar System was a chaotic one, and it isn't unreasonable to postulate that Ceres migrated from a trans-Neptune orbit to the Asteroid Belt. In that case, it would have lost its icy mantle in the process.

However, explaining its relatively circular orbit and low inclination in respect to the ecliptic plane becomes challenging if Ceres had originated from the outer Solar System and migrated inwards. It might be that Ceres' genesis is a compromise between the two theories where its rocky interior would have originated from the Asteroid Belt, while its icy material would have been delivered by outer Solar System objects such as comets. Regardless of its origin though, it seems that Ceres accumulated both water and ammonia during its formation.

In early 2017, NASA made an announcement that garnered significant media attention: Dawn had detected traces of organic compounds in Ceres' northern hemisphere, particularly near Ernutet Crater and Inamahari Crater. While scientists couldn't definitively determine the precise nature of these compounds, their spectral signatures closely resembled those of aliphatic organic compounds, akin to hydrocarbons.

What is striking with such a discovery is the fact that the aliphatic organic compounds are thought to have been formed inside the dwarf planet, implying that they had flowed onto the surface rather than being delivered in a collision with another object. Indeed, the concentration of the organics found on the surface is too high to have been transported by impacts alone, as these tend to dilute the compounds. A more plausible scenario for the origin of such organic material is through hydrothermal processing where warm water and clay minerals catalyze the production of new organic compounds. Such an interpretation not only brings to light the processes that have been active under the surface of Ceres, but it also increases the dwarf planet's potential to have in the past a habitable environment.

Indeed, what makes this discovery particularly remarkable is that the aliphatic organic compounds are thought to have originated within Ceres itself. The concentrations of these organics found on the surface are too high to have been solely transported by impacts from outer solar system objects, which usually dilute such compounds. Instead, a more plausible explanation for the origin of these organic materials is through hydrothermal processes, where warm water and clay minerals catalyze the production of new organic compounds, which later get distributed onto the surface. Indeed, this interpretation not only illuminates the potential active geological processes still occurring beneath Ceres' surface but also bolsters the dwarf planet's potential to have harbored a habitable environment in the past.

Water, Rocks, and the Potential for Life

Dawn is a complex, dynamic world that currently defies a straightforward explanation. Is it undifferentiated or differentiated? We are still not entirely sure, with some scientists stating that the protoplanet is only partially differentiated; ices and rocky material have been separated in some areas, the tell-tale sign of past heating, while other regions remain undifferentiated.

Our current understanding of Ceres' interior is the following. A solid rocky core at the center of the protoplanet does exist, however, given its density, estimated at $2.46-2.90$ g/cm^3, and the fact that Ceres might have not experienced full differentiation, it seems to be partially hydrated. More water and volatiles are to be found in the outer layers and the crust, given their densities have been calculated to be around $1.68-1.95$ g/cm^3, closer to that of water.

However, the existence of numerous craters on the surface indicates that it is a strong crust, about 30 km thick. How can the crust be similarly strong and yet have a low density? This has not been fully explained; however, it has been proposed that the mixture of rocks, ice, and salts forming the crust might also be composed of clathrate hydrates, water-based crystalline solids that trap gas molecules giving them an inherent structural strength while having a low density. More research is required to identify the crust's composition.

Gravity measurements and the lack of large craters also suggest that this ice/rock crust must lie on a weaker material. The most likely explanation is that it is an icy water mantle where pockets of liquid water exist, probably rich in ammonia and non-ice materials such as salts lowering the freezing temperature. Therefore, there is a possibility that subsurface lakes or even small regional seas might still exist today, while a subsurface ocean encircling the protoplanet is not envisaged.

Also, the distribution of minerals on the surface indicates extensive alterations, most likely due to past water/rock interaction, while the presence of ammonia point out that some of Ceres' building blocks originate from a colder environment in the outer Solar System. With the discovery of numerous brine deposits, smoother resurfaced areas, and a massive cryovolcano, Ahura Mons, reaching up to 4 km high, there is evidence that subsurface water and volatiles have been pressurized through cracks within the crust and released onto the surface, bringing with them non-ice material such as salts and minerals. Given what we saw on Enceladus in Chap. 8, it would be tempting to suggest that a substantial amount of heat is present in the icy crust. Paradoxically, it is possible that the opposite scenario is occurring; as a pocket of subsurface liquid water freezes, the ice expansion within the pocket exerts pressure on any remaining liquid brine, forcing it through cracks towards the surface.

Nevertheless, the presence of organic compounds on the surface, coupled with evidence of warm water chemistry, strongly suggests the existence of a warm liquid water environment in the past and potentially still existing today. This water environment probably constituted an ancient subsurface body of water, and maybe even a global ocean. As such, Ceres presents a unique environment where researchers have the opportunity to directly observe surfaces that have been altered by water/rock interactions, such as what might occur on the ocean floor of subsurface oceans. No other place in our Solar System is comparable.

We have seen that liquid water environments likely existed in Ceres' past and may potentially still persist today in the form of subsurface pockets. The existence of liquid water, heated by radiogenic processes and in close proximity to rocky materials, minerals, salts, and organic compounds, has prompted astrobiologists to seriously entertain the notion that microbial life may have emerged during Ceres' history and could potentially persist to this day, possibly residing deep within subsurface reservoirs of liquid water where they would be shielded from harmful cosmic rays and solar radiation. After all, with surface temperatures of up to -38 °C, one could envisage regions deep within the asteroid warmed up by radiogenic heating—as a reminder, lifeforms on Earth have been observed to survive at -25 °C.

In addition, Ceres' proximity to Earth and Mars at only 2.8 astronomical units away from the Sun does make it susceptible to biological cross contamination. In this scenario, microbial life that might have arisen on these planets during the early phases of our Solar System could have been transported via asteroid strikes to planetary objects such as Ceres, where a habitable environment could have been present under the surface. Some have also posited that this process could have unfolded in reverse, with life potentially originating on Ceres before being transported to Earth as well as possibly to early Mars or Venus, where warmer and wetter conditions prevailed. After all, objects ejected from a planetary body will usually follow the Sun's gravitational pull, drawing them inward within the Solar System. Ceres lying further out than Earth or Mars would have been ideally located to 'seed' those planets. In such a scenario, it's conceivable that we all might be Cereans.

There is still much to learn from this fascinating object and the data acquired by Dawn will keep scientists busy for many years to come. Ideally, given the unique environment Ceres has to offer, a lander should be sent to the surface for in situ measurements. However, with the scant, but real, possibility that life might exist within the asteroid, such a mission would present its own set of challenges given the severe restrictions required by planetary protection.

It is now time for us to visit two other possible ocean worlds, lying further out from Ceres. Our first stop will be at Saturn to explore its icy moon, Dione. Although at first glance Ceres and Dione might seem completely different objects, if one could imagine covering Ceres and its rock-like crust with a 100 km thick icy mantle, it would most likely bear a striking resemblance to the Dione we see today.

Dione

Discovery and Observations

With our recent exploration of Ceres and its intriguing rock-ice surface still vivid in our minds, let's now journey to Saturn's orbit. As we traverse the tenuous E ring, reminiscent of our earlier discussion in Chap. 8, we approach another candidate for an ocean world: the icy moon of Dione.

At 1122 km in diameter, Dione is somewhat larger than Ceres by only 200 km. Interestingly, in addition to Ceres, many planetary objects share a similar size to Dione: the moons Charon, Umbriel, Tethys, Ariel, and the dwarf planets Haumea, Quaoar, and Sedna are all within a 15% diameter range, although each object differs significantly in their origins and compositions.

As we make our approach, one sees that Dione has a distinctive, yet familiar surface: smooth plains, cratered terrains, ridges, and chasms, to name a few. Surprisingly, the moon's leading side—facing the direction of motion—is rather bright, contrasting with the much darker trailing side hosting a surface feature that astronomers have referred to as 'wispy terrain.' We will come back to these later.

Upon studying the surface, evidence of past geological activities affecting large areas have been found. Some scientists have even suggested that Dione might still be geologically active today, although at a much-reduced rate. Even more remarkable, it is possible that a subsurface ocean might still be present under its thick icy crust, similarly to Ganymede, Callisto, Enceladus, Titan, and Triton.

Dione was discovered in 1684 by the French-Italian astronomer Giovanni Domenico Cassini using the Paris observatory, of which Cassini was the director. In his notes, we learn that Cassini wanted to name the moon (and the three others he had found) as Sidera Lodoicea (Louisian Stars), after King Louis XIV of France, his patron, much like Galileo had done 70 years earlier when he named the four Jovian moons Sidera Medicea (Latin for Medicean stars) in honor of his patrons, the Medici family of Florence. Cassini pointed out that the names would be by themselves much more lasting monuments to the memory of the French king than those of brass and marble.

Luckily for us, the Stars of King Louis didn't catch on, and a terminology similar to the one employed for the Jovian moons was quickly adopted; Dione and her companions were known as Saturn I to VII depending on their presumed distance from the giant planet. The process of naming Saturn's moons proved to be an entertaining tale, as detailed in Chap. 7, and we had to wait until 1847, when Dione was finally named as such by William Herschel's son.

Earth-based observations of Dione in the twentieth century were limited due to the moon's small size and distant location. Nevertheless, spectrophotometric observations done in 1975 and 1976 allowed astronomers to detect the presence of ice on its surface; the first direct evidence that the moon's crust was composed of water. Further observations made shortly after thereafter revealed variations in the light reflected by the moon, known as albedo, as it orbited Saturn, indicating that its surface was not uniform.

As with the many planetary objects in the outer Solar System, the Voyager missions transformed what had previously been a point of light into a fully realized celestial body. Voyager 1 took the first close-up shots of the Dione in 1980, revealing both the leading and trailing side. Meanwhile, Voyager 2, although not as close, managed to photograph areas that had been poorly covered by Voyager 1, albeit at lower resolution. The data collected by these spacecraft enabled astronomers to calculate Dione's density with high precision, estimated at 1.48 g/cm^3. This suggests that Dione is likely composed of a substantial icy mantle overlaying a silicate core.

Upon reviewing the images returned by Voyager, it became clear to mission scientists that Dione, much like Enceladus, had experienced numerous past geological activities. Extensive surface regions resembling plains exhibit few impact craters, none bigger than 30 km in diameter. This suggests potential past resurfacing events, in contrast to ancient, heavily cratered areas with notably large craters exceeding a 100 km in diameter.

Intriguingly, the leading side showed fewer impact craters than the trailing side, contrary to what might be expected. Indeed, as a reminder, like most of the major moons in our Solar System, Dione is tidally locked to its parent planet, with one side always pointing towards the direction of motion, the leading side, and one towards the opposite, the trailing side. It was assumed that the leading side would be darker due to greater exposure to space debris and dust, analogous to a car's front windshield accumulating more dust than the rear windscreen on a dusty road. However, Dione defied expectations: its leading side is brighter. Scientists speculate that Dione was once tidally locked in the opposite direction before a significant impactor altered its spin, resulting in its current configuration.

Calculations show that due to the moon's small size, an impact large enough to form a 35 km crater could spin it in the opposite direction. Given that the surface shows many craters larger than this, it is tempting to conclude that Dione could have spun multiple times throughout the eons, although such a view is not supported by the cratering pattern observed on the surface, which can only be explained if Dione's current orientation has remained stable for billions of years. Evidently, the origin of the spin is currently not well understood, and it might be that a new idea will be proposed in the future to account for the differences in the leading and trailing side.

Another discovery that puzzled scientists was the presence of 'wispy' features on the trailing side, as seen on the images taken by the Voyagers. These elongated white fuzzy lines were thought to be the deposits from material which had vented off through linear fractures, although no sign of such activity or fissures were observed.

As numerous surface features hinted at past activity, it became clear to planetary scientists that Dione had a much warmer past. The combination of radioactive heat generated by the silicate core, as well as accretion heat trapped during the moon's formation, might have temporarily melted the ice mantle, forming a subsurface ocean which has since then frozen over. Indeed, tidal heating (described in Chaps. 1 and 6) was not considered sufficiently strong enough to warm Dione's interior given its current orbit.

The presence of chasmata, deep and elongated depressions resembling canyons, on the surface of Dione offers insights into the moon's history. These features are believed to be remnants of the moon's past, formed as its internal water mantle froze and expanded, causing the entire crust to rise. Yet, there didn't seem to be much activity today. Compared to its tiny, yet more active neighbor Enceladus, Dione seemed to be lacking in pizzazz. The arrival of the Cassini orbiter would change this view.

Cassini's Results and Dione's Habitability

The Cassini orbiter, as detailed in Chaps. 7 and 8, made just five flybys of Dione over its 13-year mission in the Saturnian system. Cassini's closest approach took place on December 12, 2011, at a distance of 99 km. The high-resolution images and instrument measurements obtained during these flybys reignited scientists' interest in this icy moon (Fig. 9.4).

The 'wispy' features initially detected by the Voyagers were found to be huge fissures in the icy crust, exposing very bright cliffs. Stretching across hundreds of kilometers of cratered surface, these fissures appeared to have formed much later in the moon's history. The most probable explanation for these giant fissures is the freezing of a former subsurface ocean. As observed with the chasmata, the freezing process would have pushed the outer layer upward, exerting stress on the crust and resulting in the formation of these enormous fissures. It is thought that cycles of freezing and melting would have contributed to the patterns we observe today, although the properties of a hypothetical ancient subsurface ocean are poorly understood. For example, there is no clear explanation as to why these cracks have only appeared on the trailing side of the moon. Whatever the cause, Dione bears witness

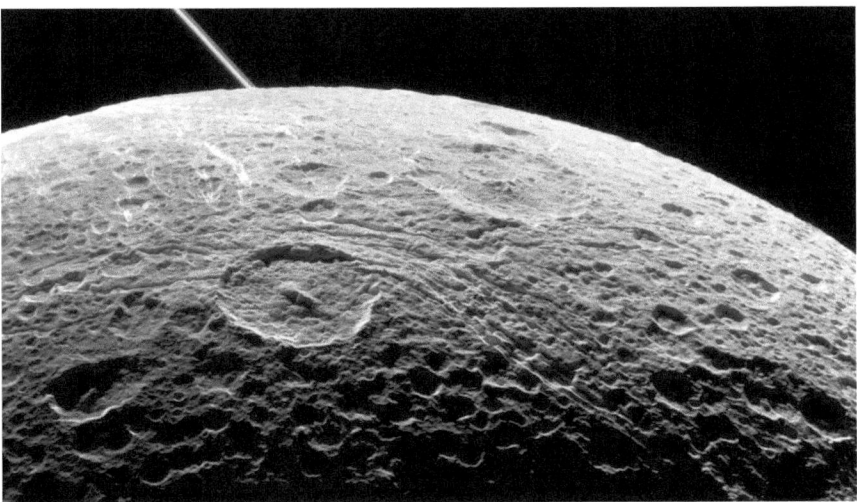

Fig. 9.4 A view of Saturn's moon Dione captured by NASA's Cassini spacecraft during a close flyby on June 16, 2015. The diagonal line in the upper left are the rings of Saturn. (Image courtesy of NASA/JPL-Caltech/Space Science Institute)

to an active past where a mantle of liquid water mixed with ammonia and other volatiles was present.

Another feature supporting such view was described in 2013 following the detailed observation of high-resolution surface images (<1 km/pixel) returned by Cassini's imaging instruments. By carefully analyzing these images, scientists discovered an area within the leading hemisphere that is entirely free of impact craters, indicative of past resurfacing, most likely due blanketed by fallen material from a nearby cryovolcano whose location has been suggested but not confirmed. In the southern pole region, scientists have also identified a surface feature that bears a striking resemblance to Enceladus' famous tiger stripes (see Chap. 8), although it is mostly inactive.

Additional evidence supporting the existence of a past subsurface ocean was discovered in the topographic anomaly of a mountain range called Janiculum Dorsa. Stretching over 800 km in a north-south direction, this ridge rises to elevations of 1–2 km. The mass of this mountain range has caused deformation and depression of the icy crust beneath it by as much as half a kilometer. This bending of the crust can be most plausibly explained if Dione had a warm subsurface ocean at the time of Janiculum Dorsa's formation.

A transient exosphere of oxygen was confirmed in April 2011 when the Cassini's magnetometer and radio and plasma wave science instrument (RPWS) detected a faint trace of ionized oxygen during a close flyby at 503 km from the surface. This exosphere is extremely thin, 5 trillion times less dense than Earth's atmosphere at sea level. Moons with exospheres composed of oxygen molecules are not uncommon—Rhea, another moon of Saturn, has one—and it is assumed that the source of the oxygen is the high-energy particles generated by a strong magnetosphere, such as Saturn's in this case, or solar radiation. These energetic particles break down the water molecules located on the top layer of surface ice, releasing oxygen molecules into the exosphere.

Remarkably, though, it seems that Dione's exosphere may also be generated by another source. In a paper published in 2007, it was revealed that measurements from RPWS detected plumes of material feeding plasma (ionized gas) in Saturn's rotating magnetosphere and concluded that it originated from within the orbits of Dione and Tethys, another icy moon of similar size. Based on what we understand about Tethys and its apparent lack of activity, scientists strongly suspect that Dione is the origin of the plasma. However, despite multiple flybys of Dione (and Tethys), Cassini did not detect any traces of a plume. This has led some scientists to speculate that the venting might be intermittent or possibly too faint to be detected by Cassini's instruments. Future observations will be necessary to verify if such venting is indeed taking place.

Considering our current understanding of Dione, it remains a possibility that there is sufficient heat within the moon's interior to support a layer of water mantle. Considering this possibility, some scientists have explored alternative methods to detect the presence of such a mantle. A 2016 study by the Royal Observatory of Belgium used gravity data from the Cassini spacecraft as it made its close approaches of the moon to explore its interior. As seen in the previous chapters, gravity data is obtained by measuring tiny variations in the velocity of a spacecraft due to the variations of the gravitational pull generated by a nearby object. Cassini continually emitting signals directed at NASA's Deep Space Network, allowing for precise measurements of its position and velocity. With this information, scientists could work out the gravitational forces acting on the orbiter at specific points around the moon. By analyzing these forces, scientists could infer differences in mass distribution, offering insights into Dione's internal structure.

According to the study, a subsurface ocean tens of kilometers thick might be located hundreds of kilometers deep within the moon. It also suggests that this subsurface ocean would rest on the silicate mantle, allowing for rock/water interactions, a crucial component for assessing the ocean's habitability. This study isn't conclusive proof that a subsurface ocean exists on Dione. In fact, most planetary scientists consider it to be very weak evidence at best. But it nevertheless suggests that there is a possibility, even a remote one, that Dione could be hosting a subsurface ocean.

Indeed, Dione most likely had an ancient subsurface ocean whose apparent freezing formed the surface features we observe today, and it might be that Dione could still have enough heat to warm up parts of its frozen interior. However, with no planned or proposed missions set out to explore Dione more thoroughly, Earth-based observations, as limiting as they can be, will be the only source of new data for decades to come. Don't hold your breath on Dione.

Based on our current understanding, is there a possibility for life to emerge and flourish if Dione indeed harbored a subsurface ocean? Some characteristics are more favorable for life than others. The fact that the liquid mantle has been predicted to be adjacent to a silicate mantle, or possibly its rocky, core does imply that it might contain dissolved elements essential for life. Furthermore, if the ocean is still warm today, it will most likely be an ancient one, formed at the same time as the moon itself estimated at around 4 billion years ago. From what we understand of life on Earth, that is plenty of time for life to arise. However, due to the limited tidal heating to warm the moon, any subsurface ocean on Dione would likely need to contain a significant amount of antifreeze compounds to remain active at low temperatures. Unfortunately, these compounds are highly toxic, making it unlikely that carbon-based lifeforms could survive in such an environment. Therefore, if Dione does indeed harbor a subsurface ocean, it would likely be cold, dark, and devoid of life.

Ariel

We now turn our attention to a planetary system we haven't encountered yet in this book: Uranus. With its collection of rings and a rich satellite system, the seventh planet in our Solar System has much to offer. In particular, a mid-sized icy moon called Ariel.

Ariel was discovered with the moon Umbriel in 1851, by William Lassell, over half a century after Uranus' two largest moons, Titania and Oberon, were discovered. Lassell, a passionate British astronomer who also happened to be a brewer by profession, also discovered Neptune's moon Triton (see Chap. 10). Although Ariel is 30% smaller than Titania and Oberon, it is still the fourth largest moon of Uranus.

Due to its apparent magnitude at 14.8 (similar to that of Pluto at its closest approach) and its proximity to Uranus, observing Ariel has always been difficult and very little was known about it. After meticulous observations of its orbit, astronomers discovered a peculiarity about Ariel: its orbit exhibits a slight eccentricity (0.0012), unlike the orbits of the other moons of Uranus. We will revisit this topic later.

Earth-based observations had also determined that Ariel's density was 1.66 g/cm^3, akin to Titania's and Oberon's, indicating a similar composition comprising half water-ice and half rocks, along with other non-ice constituents such as carbonaceous material.

We had to wait for the arrival of Voyager 2 in January 1986, to finally characterize Ariel. For a start, the peculiarities of Ariel's orbit became far more interesting for scientists when tidal heating was discovered (see previous chapters). Due to its eccentric orbit, Ariel receives heat from tidal heating as it gets squashed and pulled throughout its orbit around Uranus. Some have even called Ariel the Io of Uranus, although the moon is receiving far less energy from tidal heating than Io.

Ariel also stands out from its siblings in a remarkable way: its surface, only mapped at 35% by Voyager 2, shows evidence of resurfacing, characterized by smooth plains with few large craters. Flow-like features forming complex channel networks are also visible, hinting at large cryovolcanic events in the moon's past, where water rich in antifreeze compounds must have poured out onto the surface multiple times (Fig. 9.5). In fact, after Enceladus, Ariel displays the most active surface of any icy moon in the Saturnian and Uranian systems. Voyager 2 also discovered carbon dioxide on the moon's surface by which most likely comes from the breakdown of organic matter by bombardments of ultraviolet radiation and high energy particles.

Ariel's active past was also implied by studies showing that around 4 billion years ago, radiogenic heating from the Ariel's rocky core, estimated at 360 km in radius, was coupled with the tidal heating the moon experienced as it underwent several resonances with Umbriel and Titania. If the moon is fully differentiated, this coupling was found to be sufficient to melt a substantial part of its icy mantle that sat on top of the rocky core, creating a 'warm' subsurface ocean for hundreds of millions of years.

Ever since Voyager 2 returned such a treasure trove of data over 30 years ago, planetary scientists have wondered if the moon could still hold a subsurface ocean today. As we have seen, Ariel's orbit has a slight eccentricity and is tidally locked with Uranus (always showing the same face to the planet), subjecting it to tidal heating. Yet, with Voyager 2's only flyby to work on, we were constrained in their abilities to understand this peculiar moon.

Nevertheless, scientists are a determined bunch, and in 2023, they revisited the Voyager 2 data and made a startling discovery. In 1986, energetic ions were observed by Voyager 2's Low Energy Charged Particle (LECP) instrument in the region between Miranda and Ariel. Our understanding of Uranus'magnetosphere at the

Fig. 9.5 Ariel as seen by *Voyager 2*. Taken on January 24, 1986, this close-up shot of Ariel reveals the moon's southern hemisphere. A complex terrain of large valleys and resurfaced areas can be seen on the bottom right of the image. (Image courtesy of NASA/JPL)

time interpreted this signature as a result of the system's dynamics. New models presented show that this is not the case and suggests that a source of energetic ions is present either on Ariel or her much smaller sibling, the moon Miranda. In other words, material from one of these icy moons was venting off into space!

The possibility of Ariel supporting a plume similarly to Enceladus has been considered for many years now, with astronomers even trying to detect a hypothetical E ring around Uranus without success. On the other hand, Miranda, is a confused planetary body. It seems to have experienced traumatic events as well as some forms of tidal heating in its past. Still, Miranda is just too small to have been able to keep the heat, and it is doubtful, given our current understanding of icy worlds, that a subsurface ocean or body of water lurks under its icy shell.

It is therefore possible that Ariel is an active moon with material spewing into space. Similarly to Enceladus, could Ariel still host a subsurface ocean today despite not being in a strong resonance with other moons anymore. As with objects situated this far from the Sun, such a possibility would require a high concentration of ammonia within the ocean to keep it of freezing from the low temperatures.

The images of Ariel captured by Voyager 2 are of low resolution, covering only a portion of the moon's surface. A future mission equipped with high-resolution mapping capabilities could unveil many surprises like what we have seen on other ocean worlds such as Enceladus.

Given its warm past, the detection of material spewing out into space, and the presence of tidal heating, Ariel is, with Triton, the most promising moon within the ice giants to host a subsurface ocean. However, until a new spacecraft visits Ariel, little will be known. There is a proposal to send a flagship mission to Uranus: the Uranus Orbiter and Probe (OUP), which we will explore in Chap. 12. For the time being, we can only speculate what lies under Ariel's icy crust.

Chapter 10
Triton and Pluto

Even before Voyager 2 had sent back the first images of Triton in 1989, it was commonly accepted that Neptune's moon and Pluto—still a planet at the time—were analogs, most likely sharing a similar origin, but had somehow parted during the chaos of the early Solar System. Close observations of these objects have validated this interpretation. In this chapter, we will review both worlds, which have much alike, but are placed in different categories purely because of their location within the Solar System.

Triton

Discovery and Observations

To visit Triton, we need to travel further out from Uranus with its rich satellite system to reach the blue gas giant that is Neptune. At such a distance, 4.5 billion kilometers from Earth (or 30 AU), the Sun is just a bright jewel drowned in a sea of darkness. The expansive void surrounding Neptune may create the illusion that we've reached the outer limits of the Solar System. However, we would be mistaken. Beyond Neptune lies the Kuiper Belt, situated at 35–40 astronomical units (AU), which serves as the abode for Pluto and numerous other celestial bodies akin to it.

At 30 AU, it is now cold enough (35 K or −238 °C) for all volatile compounds to turn into ice. The most abundant ices found at this location are water (H_2O), followed by methane (CH_4), ammonia (NH_3), carbon monoxide (CO), and hydrogen cyanide (HCN). These five compounds account for 98% of the ice mass in the region. In effect, the planetary objects found here are mainly composed of rocky material and water, methane, nitrogen, and carbon monoxide.

© The Author(s), under exclusive license to Springer Nature Switzerland AG 2024 189
B. Henin, *Exploring the Ocean Worlds of Our Solar System*, Astronomers' Universe,
https://doi.org/10.1007/978-3-031-62953-2_10

The presence of these ices is the reason why Neptune and Uranus are referred to as the 'ice giants'; they hold higher concentrations of icy material than the 'gas giants' Jupiter and Saturn. When it comes to Neptune, the high temperatures generated by the planet's metallic core gives rise to layers of slushy ices composed of water, ammonia, and methane upon which rests a thick atmosphere of helium, hydrogen, and gaseous methane, giving the planet its distinctive bluish tint.

Smaller planetary bodies such as moons and dwarf planets lack the mass of planet-size objects to sustain the levels of heat required to melt the ices contained within them, so most of them are thought to be frozen solid. Although, as we shall see with Triton, this isn't always the case.

The discovery of Neptune nearly two centuries ago marked a pivotal moment in astronomy and scientific understanding. As the fourth largest planet in our Solar System, Neptune orbits the Sun at such vast distances that it eludes detection by the naked eye. Consequently, its discovery did not occur through serendipitous observations, as was common for most celestial objects up to that point. Rather, it was identified through a novel approach to exploration: mathematical predictions.

At the time, astronomers were puzzled by the peculiarities of Uranus' orbit. It was the French mathematician Urbain Le Verrier who was the first to prove through mathematics that a hypothetical planet was most likely the cause of such anomalies. Observations were done at the Berlin observatory shortly after by the German astronomer Johann Galle where he quickly discovered, on September 23, 1846, the Le Verrier's hypothetical planet at the exact location predicted by the calculations. Neptune became a sensation overnight.

Upon learning of the discovery made by French and German astronomers regarding a new planet, British astronomer John Herschel, son of William Herschel who had famously discovered Uranus six decades prior, swiftly contacted William Lassell, another esteemed British astronomer. Herschel urged Lassell to promptly commence observations aimed at identifying potential moons around Neptune, with the hope of redeeming British pride in the wake of what had become a national embarrassment in their eyes. Lassell initiated his observations the very next day.

Indeed, the British felt that Neptune's discovery was 'stolen' from them, as 2 months prior to Le Verrier presenting his work to the Académie des Sciences at Paris, a young English mathematician, John Couch Adams, had also predicted Neptune's orbit. In September 1845, Couch had contacted the director of the Cambridge observatory as well as the Astronomer Royal at Greenwich, George Biddell Airy, to entice them to start observations and find the new planet.

There is still some uncertainty surrounding the reasons why they initially disregarded Adams' predictions, but it's likely because they doubted the accuracy of his forecasts. It wasn't until Airy became aware of Le Verrier's efforts in Paris that he realized his error in not having faith in Couch's observations and failing to pursue them further. Recognizing the urgency of the situation, Airy promptly took action to make up for lost time. Thus commenced the race to discover the new planet.

By July 1846, Airy had finally managed to convince the busy and reluctant director of the Cambridge observatory, James Challis, to do a systematic search in the area of the sky where the new planet was predicted to be located in the hope of finding it first. Challis was at the time fully absorbed in the task of diligently track-ing down comets and felt that the search for a theoretical planet was not a valuable use of his time. Ironically, Le Verrier was not having much luck in persuading his compatriots as well. He had been unable to convince the French astronomers in Paris to point their powerful telescopes to the planet's predicted location, so he reached out to the more receptive Berlin observatory instead.

It might be difficult for us to comprehend Challis and the French astronomers in Paris for their lack of enthusiasm in the discovery of a new planet. It is important though to remember that, at the time, predicting the existence of a celestial object through mathematics alone was unheard of.

In fact, it was the British who first spotted Neptune. Thanks to the precise calcu-lations made by Le Verrier and Adams, Challis detected the planet shortly, thereaf-ter, recording his observations in his notebook by early August. However, he failed to recognize the significance of his findings and neglected to communicate them to Airy, likely due to a busy schedule. A month later, Galle in Berlin also located Neptune in the predicted location, and unlike Challis, he promptly published his discovery, earning him the official recognition as the discoverer of the new planet.

Upon realizing they had missed the chance to discover Neptune, the British were determined to make amends by searching for its moons. Their efforts paid off when, just 17 days after Galle's discovery of Neptune, William Lassell spotted a moon on October 10, 1846. Interestingly, Lassell, a brewer by profession, did not immedi-ately consider naming Neptune's moon. Initially referred to as "the satellite of Neptune," it was later named Triton by the French astronomer Camille Flammarion, a name that was quickly adopted by the scientific community.

Nitrogen Everywhere

The vast distances separating us from Triton, as well as its apparent proximity to Neptune (no more than 17 s of an arc), meant that observing Triton was extremely challenging. Astronomers still had very little to say 80 years after its discovery.

Things started to change in 1930 when Triton's orbit was accurately measured, revealing, to everyone's surprise, that it was retrograde compared with Neptune's spin and orbital motion. Most moons in our Solar System follow a prograde orbit, meaning they move in the same direction as their planet's rotation. Triton, however, orbits in the opposite way to Neptune rotation, the only major moon to do so. This puzzled many astronomers and led them to coming up with various interpretations.

In 1934, Issei Yamamoto, the director of the Kyoto observatory in Japan, sug-gested that a star had passed close to Neptune, forcing Triton into a retrograde orbit, and ejecting Pluto which has just been discovered a few years earlier and was thought by some to have been a moon of Neptune in the past. Another theory, pro-posed by Raymond A. Lyttleton of England in 1936, posited that a near-collision between Triton and Pluto could have resulted in Pluto's expulsion from the Neptune system and a significant alteration in Triton's orbit, causing it to orbit Neptune in the opposite direction.

Another explanation for Triton's unusual orbit was proposed: it may not have formed within the Neptune system at all but instead originated elsewhere, only to be captured by Neptune as it passed too close to the planet. However, the exact origin of Triton remained a mystery. Additionally, at the time of Triton's discovery, no new objects had been found orbiting Neptune, making it a solitary satellite system in contrast to the extensive satellite systems found around Jupiter, Saturn, and Uranus. It wasn't until in 1949, that a second moon, the considerably smaller Nereid, was detected.

In 1954, the Dutch-American astronomer Gerard Kuiper managed to estimate Triton's diameter at around 3800 km; a figure 40% higher than its actual size. This early 'erroneous' estimate had the benefit of placing Triton as one of the largest and most massive moons in the Solar System (just between Io and Callisto, the giant moons of Jupiter). Kuiper also managed to detect a mysterious reddish tint in Triton's light, hinting at the possible composition of its surface, but nothing more could be deduced at the time and Triton remained a curious oddity.

By 1978, technology had finally caught up, and methane ice (CH_4) was detected for the first time in Triton's infrared spectrum, 2 years after a similar detection had been made on Pluto, pointing once again to a similarity between these two objects. The discovery of ice on Triton meant that its surface was more reflective than what had been initially envisioned, forcing astronomers to reconfigure the moon's size to a smaller value than implied by its apparent brightness. Triton's size got closer to Pluto.

The discovery of methane on Triton provided an explanation for its slightly reddish hue, as the Sun's ultraviolet radiation can break down methane ice, transforming it into red or pink compounds. However, there's a caveat: excessive exposure to radiation or charged particles can cause the broken-down methane to turn into a black residue. This raises the possibility that if methane was indeed responsible for Triton's coloration, it would need to be continuously replenished. Situated at a considerable distance from the Sun, Triton was anticipated to be among the chilliest bodies in the Solar System. Without alternative energy sources—tidal heating had not yet been identified—there appeared to be no basis for Triton to exhibit geological activity or undergo resurfacing events that would result in the presence of fresh methane.

As the Voyagers were making their extraordinary journey past the Jovian and Saturnian systems, further infrared observations detected nitrogen (N_2) in Triton's spectra.[1]

The nitrogen measurements on Triton were so elevated that these could only be explained if the element was present not only in its gaseous state, but also in its solid or liquid state, giving scientists two distinct models for Triton's surface: one where

---

[1] *The only other place where nitrogen had been discovered on a moon of our Solar System was in Titan's atmosphere, and in much smaller quantities than what was being measured on Triton.*

nitrogen and methane ice formed a stable blanket on the surface, and one, far more exotic, where liquid nitrogen would create a vast ocean with methane icebergs floating on it—an exciting prospect at the time.

Unfortunately, prior studies conducted before Voyager 2's arrival revealed that Triton couldn't maintain the relatively high temperatures necessary for nitrogen to remain in its liquid phase (63 K or −210 °C). Consequently, the possibility of nitrogen oceans was ruled out. Astronomers were therefore confident that Voyager 2 would find, among many other features, substantial polar caps of nitrogen ice, which serves as the source of Triton's tenuous nitrogen exosphere.

Voyager 2's Discoveries

Before Voyager 2 encountered Triton in 1989, astronomers learned more about Pluto thanks to the discovery of its moon Charon (see Chap. 11) and advancements in techniques such as stellar occultations. The similarities between Pluto and Triton were striking. Both objects have a similar spectrum of visible and near-infrared light, as well as visual magnitudes arguing for bright icy surfaces. They also share nearly identical sizes, both being as large as our Moon. However, Triton is 40% denser than Pluto. It was also discovered that both objects have thin exospheres, although the composition of Pluto's exosphere was still not clearly defined at the time. Some scientists suspected that it was similar to Triton's.

Because of this, some astronomers suggested that Triton and Pluto were actually from the same family and formed in the same area of our Solar System; the so-called Kuiper Belt, a theoretical doughnut-shaped ring lying further out from Neptune. Somehow, Triton had escaped its original location and got captured by Neptune, whereas Pluto remained in its orbit. The implications of such a suggestion were profound. If the case for brotherhood was confirmed, not only does Pluto lose its status as a unique object in our Solar System, but it also argues that more Pluto-like objects should be present in this part of our Solar System, compromising Pluto's status as a planet. This also meant that learning more about Triton would improve our understanding of Pluto, and vice versa.

Because of all this, Voyager 2's planned encounter with Triton was considered to be much more than just the study of another intriguing moon in our Solar System. Instead, with Triton's visit, Voyager 2 would be unveiling of a new type of planetary object that scientists referred to as trans-Neptunian: objects formed beyond Neptune's orbit.

Intriguingly, a study led a few months before Voyager 2's highly anticipated flyby, by geophysicist David G. Jankowski from Cornell University, suggested that tidal heating could play an essential role in providing energy to Triton. Such a scenario was dependent on the obliquity of the moon's orbit in relation to the orbit's plane, and the study showed that if the moon had an obliquity of about a hundred degrees (a state referred to as the Cassini State 2), it was theoretically possible for Triton to be subjected to large amounts of tidal heating. Voyager 2's arrival a few weeks after the publication of this study would determine if Triton had the required obliquity.

And so, on August 25, 1989, Voyager 2 flew by Triton in a once-in-a-lifetime event that has not been repeated since. The good news was that since the Neptune-Triton system was the last one to be visited, mission planners could bring the spacecraft as close as possible to Triton without worrying about the angle of its outward trajectory. Triton would be, rightly so, the fitting capstone in Voyager 2's awe-inspiring journey.

On that day, the spacecraft skimmed Neptune's north pole at only 4950 km above the atmosphere; the closest approach to any planet during its 12 years journey. With this trajectory, the influence of the planet's strong gravitational pull placed Voyager 2 on a path within 38,500 km from Triton's surface. After traveling hundreds of millions of kilometers, this trajectory was remarkably precise.

As Voyager 2 zoomed towards Triton's north pole at an incredible speed of 56,000 km/h, it was too rapid for surface images to be captured. As a result, only the sunlit portion of the southern hemisphere was imaged during the flyby.

The space probe also measured the surface temperature during the close approach and confirmed that the moon was the coldest in our Solar System, with a mean temperature of 38 K (-235 °C). As a result, the surface was found to be mainly composed of ices as opposed to the nitrogen seas. Nitrogen ice was found in large quantities (55%), while water-ice (15–35%) and carbon dioxide ice made up the rest (10–20%). Surprisingly, only minute traces of methane ice (0.1%) and carbon monoxide ice (0.05%) were discovered.

The images of the southern hemisphere were particularly striking, unveiling a diverse array of surface features. Some regions appeared covered in dark streaks, while others exhibited large patches of pink ice, reflecting 70% of the sunlight that reached them. Vast plains such as Cipango Planum were found to be largely devoid of crater impacts. In fact, only 179 craters were identified within the entire set of images returned representing roughly 40% of the surface of the moon. As a rule, the smaller the number of crater impacts, the younger a planetary surface is. Some regions of Triton are estimated to be less than 50 million years old, while others are thought to be even younger, at 10 million years, making them one of the youngest surfaces in our Solar System. That Triton looked so young took everyone by surprise as this is indicative of active geology.

The oldest visible region appears to be the 'Cantaloupe' terrain: as the surface features there resemble the fruit that bears its name. Scientists are still determining its exact age of Triton's with some estimates point at a few billion years old. The challenge lies in the lack of in situ measurements and established formation theories. Most scientists consider that the surface features point towards a formation early in Triton's history (Fig. 10.1).

Within the Cantaloupe terrain, vast depressions 30–40 km in diameter suggest a process known as diapirism, where warm material rises through the icy mantle, weakening the surface strength in the process. Also, in a similar fashion to Ganymede, Europa, and Enceladus, lanes of grooved terrains representing cracks in the ice (known as sulci) were found crisscrossing the Cantaloupe terrain; the result

Fig. 10.1 Triton as seen by Voyager 2 in 1989. The 'Cantaloupe' terrain is clearly visible, as well as the south polar ice cap. (Image courtesy of NASA/Jet Propulsion Lab/U. S. Geological Survey)

of strike-slip motion along fault lines embedded within the icy crust. These features, as well as diapirisms, hint at past geological activity where heat was involved.

This interpretation was further supported by the discovery of high plateaus of volcanic origin covering most of the surface visible to the east, where multiple calderas and icy plains of lava were identified. Scientists believe that these plains are most likely composed of a mixture of ammonia and water that gushed out in as liquid or slushy ice onto the surface in the past. When ammonia is present in rich concentrations, a mixture of water-ice and ammonia can reach a melting point as low as −96 °C (177 K), making it plausible that such mixtures could be present in a liquid phase under the surface. Such exchange processes between Triton's surface and subsurface are could potentially connect a proposed ocean to its surface.

In addition, methanol can push the melting point of ammonia water-ice even lower, at around −121 °C (152 K), and can induce a viscosity similar to the ones that would have created the lava plains. One can easily imagine warm melts, such as cryolavas, of an ammonia-methanol-water mix spewing onto Triton's surface. Such a scenario implies that the moon was subjected to more heat in its past than it does today and, importantly, raises the possibility that a potentially large reservoir of liquid water and antifreeze volatiles such as methanol and ammonia was present under the surface.

Dark spots, referred to as maculae, were also found on the far eastern side of the moon. These smooth, dark patches up to a 100 km in diameter are surrounded by bright aureoles as seen at the Acapura and Zin maculaes. Reddish tints in the center of these maculaes suggest the presence of methane ice, while the brightness of the surrounding terrain implies nitrogen ice most likely mixed with some methane ice, similar to the south polar ice cap. The most likely explanation for the maculaes is a seasonal one, whereas the polar ice cap melts away in spring due to the favorable inclination of Triton and leave behind leftover patches of ice.

Even more remarkable, the images returned by Voyager 2 revealed active plumes venting material from the surface. Each plume consists of a narrow black geyser rising to 8 km in altitude, with the ejected material forming long streaks of dark clouds drifting over 150 km. Incredibly, Triton became with Earth, Io, and Enceladus one of the few bodies in our Solar System where active eruptions are detected (Fig. 10.2).

What could be driving such activity? Voyager 2 confirmed that Triton didn't have the obliquity required for it to be in a Cassini State 2, thus ruling out the possibility of it benefiting from obliquity tidal heating. Another explanation was needed.

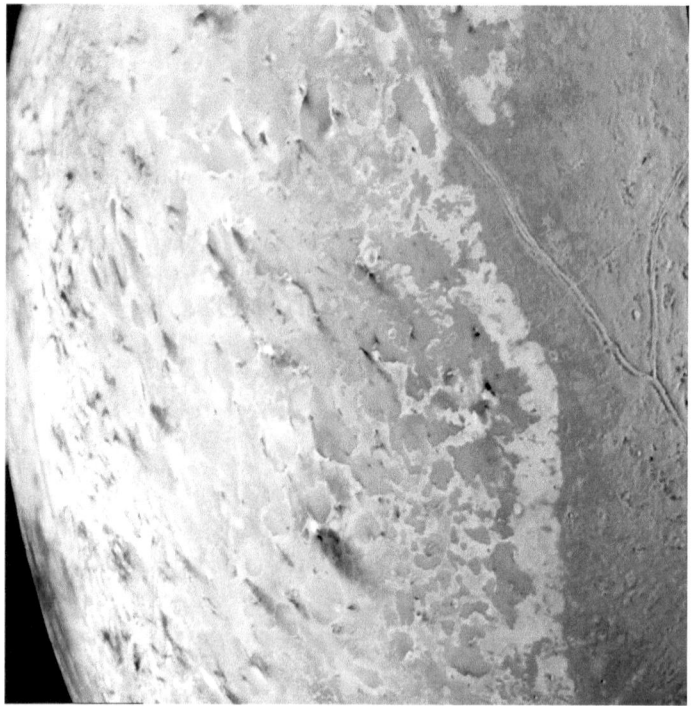

Fig. 10.2   Taken on August 25, 1989, by Voyager 2, this image of Triton's south pole reveals traces of dark plumes on the icy surface. It is possible that such vents are driven by seasonal heating of very shallow subsurface volatile deposits. Similarly, the winds transporting the particles may be seasonal winds. (Image courtesy of NASA/JPL)

Despite Triton's considerable distance from the Sun, scientists now posit that solar heating is responsible for generating these plumes, unlike the tidal heating observed on Io and Enceladus. The prevailing theory suggests that a layer of transparent nitrogen ice, approximately 1 m thick, covers the moon's south pole. Sunlight penetrates this ice layer, inducing a greenhouse effect that warms any organic material present beneath the surface. Over time, as heat accumulates, pressurized gas reservoirs form, which subsequently vent off through weaknesses in the ice layer. This expulsion carries non-ice components, such as dark organic-rich material, resulting in the distinctive black coloration of the plumes. So, contrary to Europa, the plumes observed on Triton are unrelated to the possibility of the moon hosting a subsurface ocean.

Following Voyager 2's visit, further Earth-based observations have been made of the moon's surface and atmosphere. These reveal carbon, hydrogen, oxygen, and nitrogen, and suggest the production of organic compounds, which could include similar materials that have been detected on Titan and Pluto. In 2017, changes in the moon's light curve detected by Earth-based observations and the Hubble Space Telescope indicate that transport of volatile material is taking place on its surface, although it is difficult to determine the source of these changes.

The Possibilities of a Subsurface Ocean

By precisely measuring Triton's diameter and mass, planetary scientists have calculated the moon's density at roughly 2.06 g/cm^3, indicating that the interior is composed of 65–70% rock and some metals with the rest made of ices such as water, ammonia, and methane. Studies tend to support the view that these ices form a 400-km-thick icy mantle sitting on top of the rocky (silicate) mantle.

Models suggest that radiogenic heating from the silicate mantle could potentially sustain a long-lasting subsurface ocean beneath Triton's ice shell. However, these models also indicate that this heat alone may not be adequate to explain the various surface features observed on Triton, some of which appear to be as recent as 10 million years old. As a result, the thermal history of Triton remains a mystery.

One compelling model proposed by researchers to reconcile the heat inconsistency is linked to Triton's peculiar retrograde orbit around Neptune. It goes like this; Triton was once part of a binary system—two objects orbiting each other—which passed too close to Neptune, forcing the pair to split. One member of the binary gained orbital energy and was ejected to the outer edges of our Solar System, while the other member, Triton, was captured by Neptune. Once this occurred, the new moon found itself in a highly elliptical orbit for more than a hundred million years, with the closest point to Neptune (periapsis) only five times the radius of the planet, and a semimajor axis a thousand times the radius. Neptune's strong gravitational pull would have squeezed and stretched the moon as it travelled around the planet, generating significant amounts of tidal heating in the process.

With its capture, Triton would have disrupted Neptune's original satellite system, either by colliding with bodies in it, and thus adding kinetic energy to the moon, or by disturbing their orbits and forcing them to either plunge into Neptune or to get ejected from the Neptunian system altogether. Astronomers believe this is the reason why little remains of the planet's original satellite system, which would have been like Uranus'. Today, Triton is the only large moon of Neptune, accounting for 99.5% of the total mass of Neptune's satellite system.

It has been estimated that the tidal heating experienced by Triton's capture and its subsequent orbital circularization would have produced sufficient amounts of energy to melt a significant part of the icy mantle for hundreds of millions of years and modify the moon's surface extensively. The precise timing of Triton's capture by Neptune remains uncertain, making it challenging to determine when the subsurface ocean formed. However, it is highly unlikely that Triton's capture occurred recently, while the timescale put forward for the circularization of its orbit is most likely less than 1 billion years old; although such figures are dependent on Triton's interior structure.

If a subsurface ocean was present in Triton's past, could it exist today? Despite the limited amount of tidal heating generated by the orbit's small eccentricity (0.000016), some studies showing that radiogenic heating might still be enough to sustain a subsurface ocean until now. Two studies were put forwards.

The first study, published in 2012, used a more realistic and sophisticated model of Triton's interior. It revealed that the heat generated from current tidal heating, as little as it is, is concentrated at the base of the ice mantle, especially near the poles. Contrary to radiogenic heating, which heats up the moon uniformly, tidal heating will tend to be localized at specific areas within the moon's interior. This concentration of heat prevents the icy mantle from completely freezing in that specific area, leaving a layer of liquid water near its base. There could be many areas where these layers of liquid water exist. According to the model, even a slight variation of Triton's orbit by just a few kilometers per year could sustain this layer of warmth. Additionally, the subsurface ocean would likely contain high concentrations of ammonia and methane, both of which have low melting points, although the precise composition is uncertain.

If this new model proves to be right and localized areas of liquid water exist deep within the moon, these could be sandwiched between two icy mantles. The top one would be mainly composed of nitrogen and water-ice (phase Ih) while the bottom one would be composed of high-pressure water-ice in Phase II or III; see Chap. 4 on Ganymede for further details on the different phases of water-ice. Triton is large enough for the pressures encountered at the base of the icy mantle to form high-pressure ices such as Phase II or III. Conversely, Pluto has a radius 165 km smaller, placing it just under the limit required for the formation of high-pressure ices.

Variations of this model allow the icy mantles located at the poles to be thinner due to a more pronounced tidal heating effect there, raising the possibility that liquid water from a hypothetical subsurface ocean might come into direct contact with the silicate mantle.

Another study aimed at solving Triton's energy mystery reassessed the effect of Triton's obliquity tides first predicted in 1989. The research published in 2015, and led by F. Nimmo from the Department of Earth and Planetary Sciences, Southwest Research Institute in Boulder, Colorado, used thermal evolution models to estimate Triton's heat flux and came to the conclusion that Triton's current inclination should produce obliquity tides large enough to induce convection within the thick ice shell, and subsequently form surface alterations similar to those seen by Voyager 2. The study predicts a present-day subsurface ocean of around 150 km in depth under an ice-shell of similar thickness. It also predicts that this subsurface ocean could have a temperature as high as -33 °C (240 K) thanks to antifreeze compounds such as ammonia.

What could be the composition of Triton's ocean? Given what we know of comets and the composition of other planetary bodies in the outer Solar System, scientists expect the hypothetical ocean to be rich in ammonium, sodium bicarbonate, chloride, and potentially sulfur as well.

Given what we know so far from this subsurface ocean, could life arise there? Most likely not, but we shouldn't cross Triton out yet. The low temperatures will significantly slow down biochemical reactions if not stop them altogether, while ammonia and other antifreeze compounds required to prevent the subsurface ocean from freezing at such cold temperatures will prove to be a significant hindrance for life. However, models show that the subsurface ocean will be in direct contact with the rocky core, allowing for water/rock interactions. As such, despite its extreme environment, Triton remains one of the most valuable targets in the quest for life in the Solar System.

In fact, the discoveries made by New Horizons at Pluto and Cassini at Saturn as well as recent thermal evolution models have invited a re-examination of Voyager 2's data. Planetary science relies heavily on comparative analysis and juxtaposing data from Triton with more recent observations of similar worlds like Europa, Enceladus, Titan, and Pluto, might yield new insights.

Our journey now takes us beyond Triton and its intriguing geological activity as we venture further outward towards the Kuiper Belt in the direction to Pluto, one of the most fascinating objects in our Solar System.

The Kuiper Belt

As we leave the Neptune-Triton system behind and continue to the outer part of our Solar System, we encounter a region where objects are aptly referred to trans-Neptunian objects (TNOs). Before long, we arrive at the Kuiper Belt, an expansive disk situated between 30 and 50 AU from the Sun, filled with remnants of our Solar System's formation. This region is thought to contain over 100,000 small bodies larger than a 100 km in diameter and possibly billions of icy objects between one and 20 km in diameter, exceeding the mass of the Asteroid Belt by about 20–200 times.

First theorized in the 1930s following Pluto's discovery, the Kuiper Belt was named after the renowned Dutch-American astronomer Gerard Kuiper, misleading many into believing that he was the first astronomer to predict it. As the story goes, prior to Kuiper's proposal, other astronomers had speculated about the presence of objects near Pluto's orbit. Among them were Americans Frederick Leonard and Fred Whipple, as well as the Irish astronomer Kenneth Edgeworth. Ironically, the term "Kuiper Belt" has at its source a 1951 paper written by Kuiper himself, which proposed that no objects should lie within Pluto's orbit or beyond (except for those in the Oort Cloud).

It was only in the 1980s when the Uruguayan astronomer Julio Fernández accurately predicted the existence of the Kuiper Belt using mathematical reasoning and speculated that numerous short-period comets should reside there. Because of all this, there has been much discussion within astronomy circles on the continued use of the term Kuiper Belt Objects (KBO) to define objects within the disc where Pluto is located. Some astronomers now choose to use alternate names, such as EKO in reference to Edgeworth or just TNO, although this can be misleading since TNOs refer to any objects lying past the orbit of Neptune, including the scattered disc objects or any future group of objects we haven't detected yet. For the sake of clarity and continuity, we shall use KBO in this book, even though it is now clear that Kuiper's name probably shouldn't have been chosen to define such objects in the first place.

Kuiper Belt Objects are relatively small, mainly made up of rock and icy compounds, such as nitrogen, methane, and water, and benefit from a stable orbit. Up until 1992, Pluto and its moon Charon (discovered in 1978) were the only known members of this theoretical group of objects; but all this was about to change due to the persistence and audacity of David Jewitt, a British astronomer, and Jane Luu, a Vietnamese astronomer, both working at the Massachusetts Institute of Technology at the time.

Prior to Jewitt and Luu, very few resources had been allocated to the search for KBOs. This lack of attention stemmed from the fact that KBOs were neither easily accessible targets for robotic missions, which was the main of focus of planetary exploration at the time as objects like Mars and the Jovian system were still being investigated, nor were they the primary interest of classical astronomers, who typically studied more distant celestial bodies such as stars and galaxies.

Bucking all trends, Jewitt and Luu painstakingly mapped out for 5 years the region of our Solar System where Fernández has predicted the location of the Kuiper Belt, before finally discovering, in 1992, a new Kuiper Belt object unrelated to Pluto. Designated (15760) 1992 QB1, this new KBO turned out to be small, measuring only 167 km by 108 km in size. Despite its diminutive dimensions, 1992 QB1 held significant importance in understanding the Kuiper Belt's composition and dynamics. Following this Jewitt and Luu's discovery, funding ensued and, soon enough, a new section of the Solar System was unveiled. As such, numerous KBOs have been discovered, including large objects such as Quaoar (2002), Makemake (2005), and Eris (2005).

These three KBOs were discovered largely due to the pioneering work of American astronomer Michael Brown, which made it evident that Pluto, still considered a planet at that time, was just one of many similar objects in the Kuiper Belt. This led the International Astronomical Union in 2006 to downgrade the status of Pluto to that of a "dwarf planet". While such a change proved controversial, especially in the United States where Pluto had been discovered, it was seen as a victory of scientific reasoning over historical and cultural influences. As we have seen in Chap. 9, a similar event had occurred more than 150 years ago when astronomers at the time discovered four new "planets" (Vesta, Ceres, Juno, and Pallas) orbiting between Mars and Jupiter, only to realize by 1845 that there were not planets in the classical sense and that it would be best to rename these objects as asteroids.

Since then, we have discovered thousands of KBOs and this list grows every year. Pluto is the largest member and represents a class of bodies called the Plutoids: the KBOs large enough to have attained hydrostatic equilibrium, they are symmetrically rounded into a spheroid or ellipsoid shape. Haumea and Makemake are plutoids, as was Triton prior to becoming a moon.

Pluto's Discovery and Early Observations

The discovery, classification, and later demotion of what was for more than 70 years the ninth planet of our Solar System is a classic tale in astronomy involving great wealth, serendipity, and the coming of age of a mighty nation.

The journey began in 1846 with the discovery of the planet Neptune, a feat accomplished solely through mathematical predictions based on irregularities observed in Uranus' orbit, a remarkable achievement at the time. This discovery spurred the astronomical community to explore the possibility of finding new planets using similar predictive methods.

Then, arrives Percival Lowell, a wealthy American businessman who, in addition to being a mathematician, was an enthusiastic astronomer. Based on his mathematical analysis, he became convinced that the orbits of Uranus and Neptune were influenced by the gravitational pull of a large planet situated even farther out in the Solar System. He referred to this hypothetical object as 'Planet X.'

Using his wealth, Lowell built an observatory in 1894 that bears his name and spent the rest of his life searching for the elusive planet. Ironically, Pluto was spotted by the observatory in early 1915, a year before Lowell's death. However, due to it being much fainter than Lowell's prediction, the object was disregarded as of no importance, and the search for Planet X continued.

Upon his death, Lowell's widow contested the large sums of money left behind by her husband for the observatory, and the search for Planet X ground to a halt. The dispute took over 10 years to resolve through legal battles, and in 1929, the observatory finally reopened. Clyde Tombaugh, a young amateur astronomer known for his keen eye for detail, was hired in January 1930 to continue the meticulous search for Lowell's Planet X, whose estimated brightness had been revised downward. The

search didn't take long as Tombaugh found Pluto 2 months later—in effect, redis-
covering it. The search was now on to find a suitable name for this new planet, and
Pluto was chosen following the suggestion from Venetia Phair, an 11 years old
English girl whose grandfather's brother had already recommended the names
Phobos and Deimos for the moons of Mars.

With Pluto, a Planet X had been found. Or had it? By 1931, new estimates had
dropped Pluto's mass to around one Earth mass, far lower than what Planet X had
initially been predicted to be. Additionally, by then, Pluto's orbit had been determined, and it was found to be highly inclined and eccentric, unlike the orbits of the
other planets. Pluto was an oddity.

Nevertheless, due to the technical limitations of the time, no further insights on
Pluto could be gained until 1948, when Gerard Kuiper worked out new estimations
of Pluto's mass which brought it down once more, this time to a tenth of that of
Earth's. By then, it was clear to all that Pluto wasn't the Planet X that Lowell had
been looking for as its tiny mass couldn't possibly explain the discrepancies in
Uranus' orbit.

During the following decade, significant advancements in photometry techniques
allowed astronomers to discern Pluto's characteristics more clearly. Its rotation
period was calculated to be 6.4 days, contrasting sharply with its outer planet neigh-
bors which all rotated in less than a day. In addition, variations in Pluto's light curve
indicated differences in surface albedo, suggesting the presence of a varied surface.
Furthermore, its reddish tint hinted at the possible existence of methane, a hypothesis confirmed in 1976 with the discovery of methane ice on its surface. The discovery of such a highly reflective material prompted astronomers to reassess its impact
on the planet's apparent albedo. Consequently, Pluto's mass was recalculated to be
just 1% of Earth's, further diminishing its significance among its celestial peers.

Intriguingly, such a highly reflective surface could only occur if the methane ice
was fresh, raising the possibility that Pluto's surface might be young. The methane
detection also raised the possibility of a thin methane atmosphere since the predicted temperatures, at around 40–60 K (-233 to -213 °C), were conducive to
methane vapor formation.

Pluto's biggest moon, Charon, discovered in 1978, allowed even more precise
measurement of the planet's mass through its interactions with its parent planet.
Continuing a trend, Pluto's mass was further revised downward to just 0.2% of
Earth's mass, and its diameter was now estimated to be around 4400 km.

All this implied that Pluto was far too small for its disk to be resolved through
Earth-based observations. (This would change with the Hubble Space Telescope
(HST) in the 1990s). Pluto shrank even more in 1980 when a stellar occultation
allowed astronomers to determine its diameter far more accurately than before, and
found it to be between 2300 and 2400 km in diameter; making it even smaller than
the seven largest moons of our Solar System: Io, Callisto, Ganymede, Europa, our
own Moon, Titan, and Triton. With Pluto being almost half smaller than our own
Moon, its classification as a planet began to appear uncertain. The improved accuracy of the diameter measurement enabled astronomers to deduce Pluto's density by
dividing its mass by its volume. This density was estimated to be less than 2 g/cm³,

suggesting that water was a significant component, most likely in the form of icy mantles.

Another stellar occultation 8 years later, confirmed the presence of a very thin, tenuous atmosphere, although its exact composition and temperature still proved elusive. Voyager 2's exploration of the Neptune-Triton system in 1989 marked a significant milestone for scientists studying Pluto, as both objects were believed to be closely associated. Many researchers viewed a visit to Triton as an indirect exploration of Pluto, especially since there were no planned missions to visit distant Pluto at that time.

As we saw earlier, Voyager 2's observations revealed that Triton's atmosphere was primarily composed of nitrogen, leading scientists to speculate that Pluto's atmosphere might share similar characteristics, especially considering previous detections of nitrogen on its surface.

In a further twist, more accurate measurements of Neptune's mass by Voyager 2 allowed astronomers to re-evaluate the planet's gravitational pull on Uranus. It was found that the apparent peculiarities of Uranus's orbit didn't require the existence of a Planet X and that Lowell's calculations had been wrong all along; there had never been a need for a new planet in the outer Solar System, which meant that Pluto's discovery had been the result of pure coincidence and sheer luck.

With Jewitt and Luu's discovery of 1992 QB1 shortly after Voyager 2's visit of Triton, Pluto's days as a planet were numbered.

With advancements in technology came enhanced precision in observing Pluto, leading to the detection of atmospheric methane in 1994. However, the concentrations of methane were found to be very low, less than 1%, akin to levels observed on Triton. Despite these discoveries, Earth-based observations still struggled to resolve Pluto as a discernible disk. Analyzing its reflected light was the best planetary scientists could do until the arrival of HST.

Finally, HST produced the first images of Pluto as a disk in 2002 and 2003. Advanced imaging processing techniques were used to try to define surface details, but the results were poor with Pluto's surface merely suggested. It didn't matter. The New Frontiers program, a new type of space exploration mission funded by NASA, had selected the New Horizons mission for a flyby of Pluto scheduled for July 2015.

The New Horizons Revolution

New Horizons launched successfully in 2006, ironically the same year that the IAU demoted Pluto to dwarf planet status. To shorten its journey by 3 years, New Horizon's trajectory included a slingshot (gravity assist) by Jupiter. The spacecraft carries seven instruments, including three optical instruments, two plasma instruments, a dust sensor, and a radio science receiver/radiometer; these enabled scientists to characterize, among other things, the geology, temperature, and surface composition of the planetary objects it encountered. At Jupiter, New Horizons returned an impressive set of data on the giant planet and its four Galilean moons, especially Io, where new volcanic plumes were detected.

The gravity assist it experienced made New Horizons one of the fastest space-craft ever, reaching speeds of 13.8 km/s. It still took over 9 years for it to reach Pluto in July 2015. Apart from a computer glitch that was swiftly dealt with a few days before the historic flyby, everything went according to plan and the amount of data collected on the Pluto system would prove to be exceptional. A veil had been lifted on a new world (Fig. 10.3).

Prior to the historic flyby, scientists were uncertain about what Pluto's surface would reveal. Some speculated it might resemble a uniform sphere with plentiful cratering and minimal geological activity, akin to Jupiter's moon Callisto (see Chap. 5), while others envisioned it as a simpler version of Triton showing some surface features indicative of past geological activity. Pluto's density, measured at 1.86 g/cm³, suggested the presence of significant amounts of water within the moon. Considering Pluto's mass and size, it was probable that it had undergone differentia-tion, causing heavier material to settle into a dense rocky core, while the light water rose to the surface to form an icy mantle with various other ices, including ammonia and methane.

On July 14, 2015, as New Horizons made its closest approach at 12,500 km from the dwarf planet, anticipation ran high and the data captured during the flyby stunned everyone. High-resolution images revealed a remarkably diverse surface featuring water-ice mountains cascading across plains predominantly composed of nitrogen ice intermixed with carbon monoxide and methane. Additionally, ancient cryovol-canoes, extensive resurfacing events, large plains covered in polygons, elongated

Fig. 10.3 Pluto as seen by New Horizon in 2015. The western lobe of Pluto's famous heart can be seen in this high-resolution picture taken by New Horizons. The lobe, called Sputnik Planitia, is rich in nitrogen, carbon monoxide, and methane ices. (Image courtesy of NASA/JHUAPL/SwRI)

ridges spanning hundreds of kilometers, glacial flows, and a scattering of impact craters could be seen. Regions displaying a varied color palette such as deep reds, blacks, browns, and whites juxtaposed next to each other hinting at past geological dynamism.

The mountain ranges and glacial flows on Pluto present a deceptive appearance. The frigid temperatures ranging from 35 to 40 K (-238 to -233 °C) on Pluto cause water ice to behave akin to rock on Earth, forming solid mountain ranges. Conversely, nitrogen and carbon monoxide ices, being more malleable, resemble water ice on Earth and flow across the surface as glaciers. This phenomenon leads to the formation of distinct surface features such as U-shaped valleys and moraines, which are piles of glacial debris. Moreover, because water ice is less dense than nitrogen or carbon monoxide ices, features like hills or mountains seem to "float" atop the heavier ices as they are transported across the surface.

The images also showcased frozen seas of nitrogen resting in giant basins such as the now-famous heart-shaped Sputnik Planitia. This vast basin 800 by 1050 km wide contains 168 distinctive polygonal features, also known as cells, which have an average diameter of 33 km and are thought to be 3–4 km thick. Formed by nitrogen and carbon monoxide ices, the polygons are likely the outcome of convection, a geological process characterized by the rising of heat from the interior to the surface and subsequent sinking when cooled, indicating active geological activity. These convection cells are relatively young, possibly less than 200,000 years old, making Sputnik Planitia one of the youngest areas in our Solar System. As these cells shift, fragments of mountains situated along the basin's periphery are fragmented and transported onto the plain.

Wright Mons and Piccard Mons, two possible cryovolcanoes located in the south of Sputnik Planitia. At 4 km above the surface, they are the tallest features on Pluto. Their origins are most likely due to the venting of warm pressurized nitrogen from the interior as nitrogen ices turn to vapor faster than carbon monoxide ice and are thought to be young, as they display little cratering.

Other fascinating features abound. These include the reddish north polar cap composed of irradiated methane and nitrogen ice, the penitentes (elongated blades of ice) found in the Tartarus Dorsa region, the Al-Idrisi montes mountain range of water-ice exhibiting south-facing slopes rich in methane ice, the reddish ancient terrains of the Cthulhu region layered by tens of meters of tholins (complex hydrocarbons), and the presence of a seasonal atmosphere.

Another discovery has made Pluto even more exceptional for the dwarf planet has the potential to host a subsurface ocean.

The Case for Pluto's Subsurface Ocean

Before New Horizons' visit in 2015, some scientists had already put forward the idea that a subsurface ocean might be present on Pluto today if the right conditions were right. These included the differentiation of the dwarf planet, a significant amount of radiogenic heating being generated from the rocky core, the effectiveness of the heat transfer to the surface, as well as the concentration of antifreeze compounds in the water mantle.

Scientists found that the existence of potassium within the rocky mantle would be an essential element in generating the required radiogenic heat for the melting of the icy layers. In addition, further studies suggested that the amount of potassium necessary to warm up Pluto would only have to be about a tenth of that found in meteorites hailing from the early Solar System. In theory, a subsurface ocean was possible. All that was needed now, was for the scientists to detect the tell-tale signs.

Sputnik Planitia, the heart-shaped basin, is one such sign. The basin is thought to be an impact crater formed 4 billion years ago from the collision between Pluto and a planetary object a few hundreds of kilometers in diameter—unremarkable in its ordinariness.

What makes Sputnik Planitia stand out though, is its alignment with the Pluto-Charon axis. Indeed, the basin is located directly opposite to the side facing Charon. With little chance of such coincidence occurring, estimated at only 5%, this alignment suggests that the impact crater was formed elsewhere on Pluto, possibly northwest from its present location. This is because the impact added additional mass to the basin, forcing the dwarf planet to reorient itself as large bodies in space tend to spin whichever way is easiest. Models show that Pluto most likely tilted by 60° due to the influence of Charon's tidal interactions.

We know this because Sputnik Planitia has a positive gravity anomaly; indicating that there is more gravity present in this basin than initially anticipated. This suggests the existence of high-density material beneath the surface. Only two models can explain this additional mass: either a significant amount of liquid water is located directly underneath the basin (liquid water is denser and therefore heavier than ice), or a very thick layer of nitrogen ice lies within the basin. The nitrogen ice model could be explained through the continual accumulation of ices in a local depression, which grows on itself through a positive feedback loop, although recent calculations show that the gravity anomaly would require a 40 km thick layer of nitrogen ice, which is implausible given our current understanding of Pluto's geology.

As it stands, the liquid water model fits best to explain the additional mass found under Sputnik Planitia. Here is the catch though: the presence of liquid water at the basin automatically implies that Pluto hosts a global subsurface ocean. Indeed, when the impactor hit Pluto's surface near what was then the north pole but still at the region we now call Sputnik Planitia, a counter-reaction generated an upwelling of liquid water from the subsurface ocean located directly under the impact site. The leftover crust at the impact site was thin and weakened, bulging inwards as the liquid water settled underneath it (liquid water holds less volume than ice), thus creating a basin a few kilometers deep.

In itself, the upwelling of liquid water doesn't generate the additional mass. What occurs next though does. The nitrogen in Pluto's atmosphere starts to freeze out within the basin due to increased atmospheric pressure. At the same time, nitrogen glaciers formed in the mountainous terrains adjacent to the basin began to flow inside it. Within a relatively short period of time, Sputnik Planitia becomes filled with nitrogen ice, providing the extra mass needed to alter Pluto's orientation. Such a scenario can only be explained if a global subsurface ocean of liquid water was present, ready to interact wherever an impact occurred.

Could such a subsurface ocean still be present today, or has it completely frozen out? Current models show that Pluto and other KBOs could sustain a salty subsurface ocean if it is warmed up by radioactive decay from the rocky core and richly laden with ammonia and other antifreeze compounds such as methanol or ethanol. Ammonia has been found on Charon's surface, so it is most likely present within Pluto as well. At concentrations ranging from 10% to 35%, ammonia notably lowers the freezing point of the water mantle on Pluto, rendering it syrupy in consistency, akin to honey. Given the detection of hydrocarbons such as methane and ethane, as well as more intricate molecules comprising carbon, nitrogen, hydrogen, and oxygen on Pluto's surface, it is plausible that these compounds could also be present in the hypothetical subsurface ocean. Furthermore, recent calculations propose that such an ocean might be 100–180 km thick and positioned adjacent to the rocky core.

New analysis of the images captured by New Horizons adds more credibility to the notion of a present-day subsurface ocean. For example, a study in 2023 has identified Kiladze crater, which resides northeast of Sputnik Planitia, to likely be a dry-volcano which appears to have erupted water-ice a few million years ago, contrasting with the methane and nitrogen ice covering the region.

In addition, Pluto's images do not show signs of compression which should take place if the subsurface ocean would have cooled down and frozen with time. As water expands during freezing, Pluto's surface should have stretched, generating recognizable surface features. None have been identified on the dwarf planet so far.

Furthermore, models have shown that the dark ripple-like features seen on the opposite side of Sputnik Planitia, Pluto's far side which is only available in low-resolution images, can be explained if the impact energy that created Sputnik Planitia would have sent shock waves throughout the dwarf planet, focusing the energy at the exact opposite point of the impact. These models show that this is only possible if a mantle of liquid water exists.

Nonetheless, a study published in April 2024 has presented an alternative interpretation regarding the surface characteristics of Pluto, suggesting that a subsurface ocean may not be necessary to account for the gravitational anomaly observed in the heart-shaped basin.

The researchers in the article have utilized advanced hydrodynamic simulations to propose that a collision involving a low-velocity ice-rock object approximately 730 km in size could account for the alignment of the Pluto-Charon axis and the formation of Sputnik Planitia's morphology. This explanation does not necessitate the presence of a liquid water layer within the mantle although more research is required.

If Pluto does have a subsurface ocean today, could life have arisen there? Although extremophiles might be able to find a way to survive such cold temperatures or high concentrations of salt, it is the noxious levels of ammonia that contribute the most in preventing the emergence of life as we know it. If life could arise in Pluto's exotic ocean, it would be very different from life on Earth, alien.

Regrettably, Pluto's distance of 40 AU makes new missions to the dwarf planet unlikely in the near future. Nonetheless, preliminary proposals have surfaced for a Pluto orbiter mission, which would utilize an ion engine similar to the one employed by the Dawn spacecraft. This proposed mission could conduct a comprehensive study of Pluto and its moons over a span of 4–5 years, leveraging gravity assists from Charon to propel itself from one celestial object to another. There have even been calls for such a mission to be launched by 2030, marking a historic milestone as it coincides with the hundredth anniversary of Pluto's discovery, although it seems very unlikely at the time of writing.

In the forthcoming chapter, we will explore other planetary objects that may host a subsurface ocean.

Chapter 11
The Possible Others

In the previous seven chapters, we have investigated eight icy satellites and two dwarf planets within our Solar System. Five of those have already been confirmed as ocean worlds, whereas five are very likely to host a subsurface ocean, or subsurface sea in the case of Ceres, although conclusive evidence is still pending. That our Solar System can host so many planetary objects capable of sustaining large amounts of liquid water is remarkable. Still, there is potential for more. In this chapter, we will consider the possibility of subsurface oceans in other icy satellites and trans-Neptunian objects (TNOs) for which we currently have little observational data. Some of these objects might be confirmed as hosting past or present subsurface oceans in the coming decades.

This chapter begins by revisiting the Saturnian system. Over the course of Cassini's 13-year mission around the ringed planet, extensive data were collected on Titan (refer to Chap. 7) and Enceladus (refer to Chap. 8), leading to the discovery of subsurface oceans beneath both moons. While Dione (refer to Chap. 9) shows promise, evidence for a subsurface ocean has proven to be elusive. However, Rhea, another mid-sized icy moon orbiting Saturn, emerges as a potential candidate for an ocean world.

Rhea

Discovered by Cassini in 1672, Rhea suffers from a lack of personality despite being Saturn's second largest moon after Titan. Its leading hemisphere features numerous ancient craters, a common feature for a moon orbiting Saturn. Conversely, its trailing side exhibits younger regions, hinting at previous resurfacing events and

© The Author(s), under exclusive license to Springer Nature Switzerland AG 2024   209
B. Henin, *Exploring the Ocean Worlds of Our Solar System*, Astronomers' Universe,
https://doi.org/10.1007/978-3-031-62953-2_11

fractured terrain, although these characteristics are less prominent compared to other moons. Consequently, Rhea fades in comparison with Dione and Enceladus.[1]

Cassini made only five close flybys of Rhea throughout its 13-year mission around Saturn, during which many of the moon's characteristics were refined. The mean density has been precisely measured at 1.23 g/cm^3, implying that Rhea is mainly composed of water with a quarter rocky material and three-quarters water-ice. Much uncertainty remains as to how these are distributed within its interior as differentiation might not have occurred due to the small amount of rocky material available to generate radiogenic heating and the minimal tidal heating present. Either distinct layers of ice and rock exist within Rhea, or these are instead distributed homogeneously throughout. Multiple studies on this topic have reached different conclusions.

This uncertainty prevents scientists from being confident in their interpretation on how Rhea was formed. It could either have coalesced rapidly, trapping a significant amount of accretion heat in its interior that would allow it to differentiate, or it could have formed slowly, like Jupiter's moon Callisto, thus keeping its interior as a homogenous mass or at least partially differentiated. If Rhea is fully differentiated, it has been estimated that the rocky/iron core would be large enough, about 350 km in radius, to produce enough radiogenic heat in addition to the leftover accretion heat to melt the adjacent layer of water-ice creating a liquid mantle in the process. A partially differentiated interior would have a bigger core region with poorly defined boundaries consisting of a rock and ice mix. This could, in theory, allow for pockets of ice to melt, due to the heat generated from radiogenic heating, in a similar manner to Ceres.

Surface features such as young terrains and chasmata do show that there has been past internal activity, although there is no clear indication as to what precisely could have driven this. It might be that, like Dione, a past subsurface ocean was responsible for these features. Whether a liquid mantle still exists under Rhea's ice shell is difficult to determine; however, the latest studies seem to refute such a possibility. More studies are required.

Apart from Rhea, Iapetus and Tethys are two other mid-size icy satellites orbiting Saturn that have been found to be completely frozen solid, removing all possibility of subsurface oceans occurring. Therefore, the Saturnian system might potentially host four ocean worlds: Titan, Enceladus, Dione, and Rhea.

We shall now leave Saturn and its satellite system to meet other potential ocean worlds orbiting another exciting planet, Uranus.

---

[1] It was thought for a while that Rhea might have a ring system, a first for a satellite, but this has now been discarded due to a lack of evidence.

Uranus' Moons

Uranus, situated midway between Saturn and Neptune, ranks as the third largest planet in our Solar System. Frequently overlooked, it boasts a plethora of features, including a ring system, an intriguing atmosphere composed of ices and gases, and a diverse array of moons teeming with mysteries and potential discoveries.

Of the 28 icy moons that form Uranus' satellite system, in addition to Ariel which we have already explored in the previous chapter, three other mid-sized moons are of great interest: Titania, Oberon, and Umbriel.

Uranus' moons typically exhibit darker tones compared to Saturn's icy moons, a characteristic attributed to the significant presence of non-ice components during their formation. This phenomenon also accounts for their relatively larger rocky cores in proportion to their sizes. An important characteristic due to the substantial radiogenic heating these cores will generate.

Voyager 2's brief flyby in 1986 remains the sole opportunity we've had to observe Uranus and its moons up close, as no spacecraft has ventured to the Uranian system since. Due to the unique tilt of Uranus and its moons, which are tilted by a staggering 97.7 degrees, akin to lying on their side, Voyager 2 was forced to navigate perpendicular to the system rather than parallel, like a dart flying through a dartboard. Such a trajectory limited the observation time Voyager 2's had for each moon.

Since then, Earth-based observations have continued to provide valuable insights into these moons. Among Uranus' numerous moons, Titania and Oberon stand out—with Ariel—as the most promising candidates for hosting past or present subsurface oceans.

Titania

Discovered by Herschel 6 years after his discovery of Uranus, Titania holds the distinction of being the largest and most massive among all of Uranus' satellites. Similar in size to Saturn's moon Rhea, little was known of Titania until the arrival of Voyager 2 which returned images of a relatively young surface, although only 40% of the moon has been imaged. Various surface features have been found on the icy shell such as chasmatas (canyons) and rupes (scarps). These suggest that the moon cooled down during its past, and had its liquid mantle frozen, although we are not sure.

So far, only two compounds have been detected on the surface: water-ice and carbon dioxide. The latter could be the by-product of the decomposition of organic material under the constant bombardment of ultraviolet radiation and charged particles from space. This might also explain the reddish tint detected in some areas of the moon, hinting at the breakdown of carbon compounds.

The moon's density, calculated at 1.71 g/cm^3 (much denser than Dione or Rhea), suggests that it is composed equally of rock and ice. We do not know if Titania's interior has been differentiated, as our understanding of the moon's history is very poor. Nevertheless, if differentiation has occurred, and if ammonia or methane are present in sufficient quantities to act as antifreeze, the moon could, in all likelihood, host a subsurface ocean thanks in part to the radiogenic heat generated by the abundant rocky material.

Recent models suggest that such a subsurface ocean would be directly adjacent to the large rocky core and be as thick as 50 km. The ocean would be very cold, though, with some models predicting temperatures as low as 190 K (-83 °C), a seemingly insurmountable challenge for life. Additionally, the high concentrations of ammonia and other antifreeze compounds required for liquid water to exist at such low temperatures would also impede life.

A future mission to Titania is required if we want to know more about the moon's interior. One way such a mission could ascertain the existence of a subsurface ocean would be to measure its influence on Uranus' magnetic field, similar to how Europa's subsurface ocean was detected; see Chap. 6 for further details. There are proposals to launch an orbiter to visit Uranus and its moons by the mid-2040s. Until then, though, we can only speculate about Titania and its potential for hosting a subsurface ocean today.

Oberon

Slightly smaller and less massive than Titania, Oberon shares a similar story. Discovered by Herschel on the same day as Titania, Oberon was finally brought to light almost 200 years later thanks to Voyager 2's flyby. The spacecraft imaged 40% of the moon, revealing a more heavily cratered and redder surface than its larger sibling. This is in large part due to Oberon being the outermost large moon of the Uranian system where it encounters far more small objects such as small irregular satellites roaming the edges of the satellite system. Surface features such as cracks and canyons also appear to have been formed as the moon expanded due to the freezing of a liquid mantle, but these fissures are less prominent than on Titania.

Oberon's density is within the same range as Titania, at 1.63 g/cm^3, leading scientists to believe that it should also have a significant rocky core at approximately 480 km in radius. If differentiation has occurred, recent models show that radiogenic heat is sufficient to melt the icy mantle if the water is rich in ammonia. Such a liquid mantle would be formed next to the rocky core, allowing for water/rock interactions, although given the low temperatures and presence of antifreeze compounds such as ammonia, we end up with the same conclusions as we did with Titania: the subsurface is unlikely to be habitable. Whether such subsurface ocean does exist is something a future mission will have to confirm.

At present, we can only speculate and wonder if Uranus holds, with its three moons: Ariel, Titania, and Oberon, a collection of ocean worlds similar to Jupiter and Saturn.

The Centaurs

Before we finally depart from the realm of the giant planets and explore the possible ocean worlds inhabiting the Kuiper Belt and beyond, it is worth mentioning that there is a class of Solar System objects that hasn't been introduced so far: the centaurs. These small objects can be found wandering between the orbits of Jupiter and Neptune and are known to have unstable orbits. They only stay for a few million years within the outer planets region before their trajectories take them somewhere else, either out of the Solar System or closer to the Sun.

Overall, not much is known of the centaurs as they are difficult to observe, being relatively small and dark in color. Nevertheless, it is widely accepted that these objects didn't form at their present location but are instead small TNOs that got dislodged from their original orbit due to an unfortunate encounter. Given that some TNOs are thought to be ocean worlds, could centaurs be potential candidates as well?

It is highly unlikely. The biggest centaur we have discovered so far is Chariklo with a radius of only 124 km, making it smaller than Saturn's moon Mimas, the smallest spherical known object at 198 km in radius. At these truly small sizes, differentiation can't occur as not enough heat is trapped to initiate the melting process. Therefore, it is not expected for these objects to host subsurface bodies of water. Although current estimates suggest that there are approximately 40,000 centaurs in our Solar System with diameters exceeding 1 km, and only 500 have been identified thus far, any centaur larger than Chariklo should have been detected by now.

Centaurs are just too small to be considered ocean world candidates, but by being much closer to the Sun than most TNOs, they are still brighter and more accessible. Given their origin, by further studying centaurs, we can better understand TNOs whose distance make them truly inaccessible for now.

More Kuiper Belt Objects

Located within the outer edge of the planetary system where the gas densities are too low and the accretion timescale too high for a single dominant planet to form, small planetary objects reside in a doughnut-shaped disk named the Kuiper Belt.

Due to their diminutive sizes and the immense distances that separate them from us, Kuiper Belt Objects (KBOs) have proven exceedingly challenging to detect, except for Pluto itself. For an introduction to the Kuiper Belt and Pluto's discovery, please see Chap. 10.

Triton, thought to have originated from the Kuiper Belt before being captured by Neptune, is the largest among them all. And until New Horizons visited Pluto in 2015, Triton was the only KBO that had been observed up close thanks to Voyager 2.

With the discovery of new KBOs in the 1990s, their remote locations presented significant challenges to researchers, despite the utilization of the most advanced telescopes on Earth. Spectroscopic and photometric studies produced results of moderate quality and were challenging to interpret.

In the last two decades, the discovery of larger KBOs has enhanced our understanding of these enigmatic objects. Thus, we have gained a better understanding of the processes that can influence their surface compositions. We will begin our exploration of KBOs that may have harbored a significant subsurface ocean in the past or could potentially do so now with Pluto's moon Charon, also visited by New Horizons.

Charon

Many planetary scientists were intrigued when, in 2015, the New Horizons spacecraft returned images from Charon, Pluto's largest moon (Fig. 11.1). Little was known of the moon at the time, which can be pronounced 'Karon' or 'Sharon.' The latter is preferred by its discoverer, the U.S. Naval Observatory astronomer James Christy, who named the moon in 1978 after his wife, Charlene.

Fig. 11.1 A beautiful composite image of Charon taken by New Horizons on July 14, 2015. One of the most striking features is the reddish north polar region, informally named Mordor Macula. (Image courtesy of NASA/JHUAPL/SwRI)

Before the New Horizons' flyby, many suspected Charon to be a frozen world unchanged since its inception and pockmarked with countless craters. As is often the case in planetary science, the moon turned out to be rather different. High-resolution images sent back by New Horizons revealed a diversity of geological features as well as the eye-catching Mordor Macula and Charon's north pole, coated in red; the result of the moon's gravity drawing in Pluto's atmosphere which freezes and falls as nitrogen and methane ice onto Charon's surface. Subsequently, solar radiation breaks down the methane, giving it the distinctive red coloring.

Even more intriguingly, smooth plains in the southern hemisphere, such as Vulcan Planum, exhibit a relatively youthful surface, suggesting previous resurfacing events likely attributed to cryovolcanism. This phenomenon involves the eruption of "warm" liquid water abundant in ammonia and other volatiles onto the surface. In support of this interpretation, a team of scientists using the Earth-based Gemini observatory in 2007 detected traces of ammonia hydrates and water crystals in Charon's spectra. The presence of water-ice in its crystalline form hints at recent resurfacing events as this type of ice usually decays into amorphous ice after a few tens of thousands of years due to solar ultraviolet radiation and cosmic ray bombardment. In other words, there is fresh ice on Charon.

The discovery of ammonia hydrates is also telling as it suggests that this anti-freeze compound is most likely present within the icy mantle below. With the right conditions, these compounds could contribute to the existence of pockets of water-ammonia slush under the surface. During its flyby, New Horizons did not detect any indication of active resurfacing events. However, this absence of activity is not unexpected, as such events are believed to occur only once every tens of thousands of years.

Nevertheless, Charon's surface features reveal deep fissures and rifts, suggesting that the moon had a subsurface liquid mantle in its past. Indeed, while slightly less dense than Pluto at 1.70 g/cm³, Charon still comprises a significant proportion of rock to ice, especially when compared to the icy moons of Saturn. Radiogenic heating at the core likely melted the bottom of the icy crust, forming a subsurface ocean. Unfortunately, the moon's size wasn't big enough to sustain the primordial heat, and Charon cooled with time, freezing the subsurface ocean with it. As frozen water expanded, the crust was lifted up, giving us the surface features we see today.

It does appear, however, that considering the relatively recent resurfacing events that may have been observed by New Horizons, there could potentially still be sufficient heat to sustain very small pockets of water-ammonia slush beneath the surface.

Makemake, Quaoar, Salacia, Orcus, and 2002 MS4

That the three large Kuiper Belt Objects we have visited—Triton (previously a KBO), Pluto, and Charon—hold much potential as past or present ocean worlds is not as surprising as it might seem at first. Ultimately, KBOs are mainly composed of ices and rocks, and the conventional thinking goes that the more massive objects are, the denser they should get as gravity causes them to compact. If a Kuiper Belt Object possesses enough mass to retain sufficient heat from accretion and radioactive decay, it is likely that differentiation has taken place. With sufficient heat and suitable composition of interior ices, coupled with significant amounts of antifreeze compounds, large sections of the water ice mantle might melt, leading to the formation of a subsurface ocean that may endure to the present day.

Unfortunately, there is no easy way to determine at what size/mass limit differentiation occurs, as this will depend on many variable factors, such as the distance from the Sun when the object was formed, the ratio of ice to rock, how fast accretion has occurred or the quantity of radioactive material aggregated such as Al-26.

For example, we have come to realize that KBOs with a similar radius share very different properties, implying that although it might be theoretically possible for KBOs of a certain size to undergo differentiation and host a subsurface ocean in the past, this needs to be reviewed on a case-by-case basis. Although, planetary objects with a radius greater than 1000 km such as Triton and Pluto are massive enough to have had a subsurface ocean in the past and might still do today. At 780 km in radius, Haumea might also be considered part of this group. However, its highly elongated ellipsoidal shape believed to result from a tumultuous history likely caused by an impact that fragmented part of its icy mantle complicates the assessment of whether the conditions were suitable for the formation of a subsurface ocean.

Smaller mid-size KBOs, ranging from 200 to 500 km in radius, such as Makemake, Quaoar, Salacia, Orcus, and 2002 MS4 are still large enough to have undergone differentiation and retained some internal heat.

As such, understanding a KBOs density is crucial for gaining insight into its potential for hosting a subsurface ocean. Consider two KBOs Salacia and Orcus. Salacia boasts a larger radius (450 km) compared to Orcus (400 km), theoretically increasing the likelihood of harboring a subsurface ocean. However, Orcus exhibits greater density (2.3 g/cm^3) and surface ice presence (higher albedo), suggesting differentiation. Conversely, Salacia's low surface ice content (low albedo) and lower density (1.1 g/cm^3) indicate potential non-differentiation. This variance in properties among similar-sized objects may stem from Salacia forming later than Orcus, accumulating more icy material akin to comets rather than denser rocky material. Another example, the much smaller KBO 2002 UX25 has a radius of 105 km and has been found to be mainly porous due to its low density.

Without the capability to visit these objects directly, our best approach to understand them is to characterize their surface chemistry through spectroscopic analysis, although this method may not provide a complete representation of the entire body. Makemake's surface seems to be dominated by methane ice, while Orcus and Quaoar show strong bands of crystalline water-ice in their spectra, which some have interpreted as evidence for recent resurfacing events. Not everyone is convinced by this interpretation, though, so the presence of crystalline water-ice remains a mystery. Ammonia is most likely present in these objects as well, however, since this chemical is complicated to detect, it hasn't shown up in the spectra yet.

In 2023, the James Webb Space Telescope (JWST) surveyed multiple Trans-Neptunian Objects and identified complex organic compounds on Quaoar, resulting from the irradiation of methane (CH_4). Ethane, another hydrocarbon molecule, water (H_2O) and carbon dioxide (CO_2) were also detected, providing insights into the conditions prevalent on the planetary body. Makemake was also recently observed by JWST. These observations not only revealed that its surface shows signs of complexity, but may actually harbor a subsurface ocean today.

As such, the JWST observations focused on methane, and in particular its D/H ratio, the deuterium (heavy hydrogen, D) to hydrogen (H) ratio found in methane. Deuterium is thought to have originated from the Big Bang, while hydrogen is the most prevalent nucleus in the universe. The D/H ratio observed on a celestial body provides insights into the origin, geological evolution, and formation processes of hydrogen-containing compounds. It is like a window of what occurs within the dwarf planet.

The D/H ratio measured by JWST on Makemake points to geochemical origins for methane produced within their deep interior. The latest models suggest that methane must have originated from the high temperatures found in the rocky cores of these dwarf planets. It is thus possible that Makemake still hosts cryovolcanic processes within its rocky core, therefore delivering the methane to the surface. The dwarf planet will probably exhibit evidence of recent resurfacing such as seen on Pluto, although no mission is planned to visit this distant world anytime soon. The crucial point here is that the discovery of newly formed methane, along with the suggested resurfacing events and geochemical activity, indicates that there might likely be enough energy within Makemake's interior to melt a substantial portion of the icy mantle, potentially leading to the formation of a subsurface ocean.

Given to our limited knowledge of these distant objects, it is very challenging to determine whether any of these objects harbored a subsurface ocean in the past, let alone today. As scientists refine their models and conduct additional observations of KBOs, it is probable that more intricate surface features will be revealed. And with these, we might gain further insights into their ability to host subsurface oceans.

Scattered Disk Objects

Further out from the Kuiper Belt lies another group of objects made of rocks and ices that form a large irregular disk referred to as the Scattered Disk. In contrast to KBOs, which maintain stable orbits around the Sun, scattered disk objects (SDOs) host bodies with significantly more irregular orbits. This irregularity arises from Neptune's influence during the early stages of the Solar System's formation, leading planetary scientists to theorize that SDOs serve as the origin of short-period comets, whereas long-period comets originate from the Oort Cloud. Most SDOs follow highly elliptical orbits, with some extending up to a hundred AU, marking them as the most distant objects observed within our Solar System. However, their eccentric orbits can also bring them into proximity with Neptune's orbit during their closest approach.

The biggest SDO detected so far is Eris at 1163 km in radius, slightly smaller than Pluto, yet 27% more massive, implying a rockier composition. In fact, Eris is the ninth most massive planetary object directly orbiting the Sun after the planet Mercury. Pluto is the tenth. It is therefore no surprise that Eris' discovery in 2005 is one of the events that led to Pluto's demotion to a dwarf planet.

Our understanding of Eris remains relatively restricted as no space probe has been sent to investigate it directly. However, we have been able to measure certain characteristics, such as its surface temperature ranging between 30 and 56 K (-243 and $-217°C$), and its albedo being considerably brighter than Pluto or Triton. Unlike these celestial bodies, Eris lacks their distinctive reddish hue attributed to the decomposition of organic compounds known as tholins. The detection of methane in Eris'spectra suggests that it has uniformly condensed across its surface, with the ice effectively reflecting a significant portion of sunlight. Like Makemake, JWST observed methane on Eris with D/H ratios indicative of geological activity. Building the case that Eris is an active world, with a perihelion of about 37.9 AU, it might be possible for methane ice to sublimate into gas and form a light atmosphere, which in turn might slowly escape into space. If this is the case, the methane would need to be replenished by an active geological process.

Furthermore, as Eris is denser than Pluto, models have shown that the radiogenic heating produced by its larger rocky core could be capable of sustaining an internal ocean of liquid water, provided that it is also rich in ammonia. Unfortunately, it would take at least 25 years for a spacecraft to reach Eris using conventional methods; we will most likely never know what lies beneath Eris' surface.

Another intriguing SDO is a contender for the dwarf planet category, Gonggong. At 751 km in radius (similar to Rhea), this object is the third largest known TNO after Pluto and Eris, making it slightly bigger than Makemake—some scientists consider Haumea to be bigger than Gonggong, even though its ellipsoid shape gives a smaller volume. Spectra analysis of Gonggong has detected water-ice and methane that, due to its highly eccentric orbit, might form a tenuous atmosphere at its closest approach to the Sun at 33 AU in a few hundred years from now; it is currently at 87.5 AU and speeding towards aphelion at 101 AU.

Gonggong's density is at 1.74 g/cm³. Given its size, it is likely to be differentiated with a rocky core and an icy crust made of water-ice and methane, similar to Pluto and Eris. Recent models have shown Gonggong to be capable of sustaining a subsurface ocean as long as antifreeze compounds are present there. JWST also observed Gonggong recently and found ethane features as well as water and carbon dioxide.

That leads us to the final Solar System object, which will be reviewed as a potential ocean world: the enigmatic Sedna. Discovered in 2003, Sedna's orbit is an extremely eccentric one, taking it from 76 AU at its closest to a whopping 936 AUs, far greater than most SDOs. The orbit is so elongated that Sedna takes 11,400 years to go around the Sun. It is intriguing, since Solar System objects don't start off in a highly eccentric orbit. Instead, they tend to have a circular orbit during their formation, and this is most likely how Sedna started, somewhere around 75 AU from the Sun. However, its orbit was later stretched into its present course by an unknown external agent.

Proposals for the agent include an undiscovered planet lying at 2000 AU (the hypothetical "Planet 9"), the effect of a passing star early on in our Solar System's history, or even more radical, that Sedna is a captured extrasolar planetary object. Given the scope of all these plausible explanations, understanding the cause for Sedna's unusual orbit might be the key to understanding our Solar System's origin and evolution.

Sedna's peculiar orbit presents a challenge for some astronomers. They question its classification within the Scattered Disk category, as Neptune's influence is minimal at such distant reaches from the Sun, and most SDOs don't exhibit orbits as highly eccentric as Sedna's. Some speculate that Sedna might actually belong to the hypothetical inner Oort Cloud, or that the Scattered Disk extends further out than previously thought. Consequently, Sedna could represent a new category of planetary objects termed Sednoids, which have a perihelion well beyond 47 AU. Sedna could also be considered as being part of the inner Oort Cloud objects. Regardless of how Sedna should be classified, the reality remains that our knowledge of this inaccessible region of the Solar System is currently very limited.

We do know, thanks to Earth-based observations, that Sedna has a radius of around 498 km, making it slightly bigger than Ceres (see Chap. 9). However, its mass is undetermined as it has no moon which could enable us to measure its mass.

Intriguingly, Sedna's reflected light is one of the reddest in the Solar System, suggesting that hydrocarbons such as methane have decayed into tholins due to long exposures of UV radiation and solar particles. This, in turn, implies that resurfacing activities such as icy cryovolcanoes are not present. JWST also detected a large number of absorption features from ethane (C_2H_6), as well as small amounts acetylene (C_2H_2), ethylene (C_2H_4), and possibly carbon dioxide.

Similar to Gonggong and Eris, Sedna theoretically could have harbored a subsurface ocean in its past if its rocky mantle is sizable and the icy mantle contains sufficient quantities of antifreeze compounds. There is also a possibility that Sedna could still host a subsurface ocean at present.

Like all distant Solar System objects, unless a spacecraft is dispatched to visit Sedna, it will remain an enigma. Given its orbital period of over 11,000 years, the years 2075–2076 represent our sole realistic opportunity to explore Sedna up close as it will be at its closest approach to the Sun at 76 AU.

With many more objects of similar size likely populating this remote region of our Solar System, the potential for discovering additional planetary bodies capable of hosting a subsurface ocean is poised to expand. As we conclude this journey through our Solar System with Sedna, the farthest of the ocean world candidates, we extend gratitude to our robotic emissaries: Pioneer 10 and 11, Voyagers 1 and 2, Galileo, Cassini-Huygens, New Horizons, Juno, and Dawn, along with Earth-based ground and space telescopes. Their discoveries within the last 35 years have revolutionized planetary science and astrobiology.

So, what lies ahead in the next 35 years? It is now time for us to gradually shift our focus back towards Earth. In Chap. 12, we will explore the ongoing initiatives being devised for the exploration of the ocean worlds and try to anticipate what discoveries might be made in the future.

Part IV
Future Missions to the Ocean Worlds

Certainly one of the most enthralling things about human life is the recognition that we live in what, for practical purposes, is a universe without bounds. (James Van Allen)

We are living through a new age of space exploration as the momentum for the investigation of ocean worlds is growing among the public and scientists. In this fourth and last part, we cover the confirmed missions currently being put in place by space agencies as well as proposed missions waiting to be selected or in need of further development.

Chapter 12
Confirmed and Proposed Missions to the Ocean Worlds

A New Frontier

The last decades of space exploration have been remarkable. No less than five ocean worlds have been identified in our Solar System, and, as we have seen throughout the previous chapters, there is potential for more to be confirmed. Not only is the volume of liquid water locked in the moons of Jupiter and Saturn vastly greater than what we have on Earth, the fact that at least two of the subsurface oceans are in direct contact with rocks is remarkable. This marks a departure from our earlier perceptions of the Solar System. As discussed in Chap. 6, it was in 1971 when astronomer John S. Lewis initially suggested the possibility of liquid water existing beneath the icy surfaces of small planetary objects like moons. While this concept garnered little support initially, contemporary scientists now acknowledge that rather than being the exception, such water-rich habitats may be commonplace throughout our Solar System.

Such a paradigm shift would not have been possible without significant investments which allowed us to send a few intrepid spacecraft to the outer planets: two Pioneers, two Voyagers, Galileo, Cassini, and New Horizons. And yet, despite the substantial progress made in recent decades, our understanding of the ocean worlds remains constrained, primarily owing to the immense distances involved. While more than 50 robotic probes, orbiters, landers, and rovers have been sent to Mars over the past half-century, only a few spacecraft have ever flew by Europa and Enceladus.

This knowledge gap has been recognized by the major space agencies and the lawmakers who ultimately fund them, and we are now experiencing a new age in planetary exploration where, in addition to building spacecraft specifically designed to investigate these ocean worlds, the agencies are also investing resources into maturing the technologies required for future missions to these worlds.

© The Author(s), under exclusive license to Springer Nature Switzerland AG 2024   223
B. Henin, *Exploring the Ocean Worlds of Our Solar System*, Astronomers' Universe,
https://doi.org/10.1007/978-3-031-62953-2_12

As such, both NASA and ESA have a flagship mission, each planned to visit the ocean worlds of Jupiter in the coming decade; NASA with Europa Clipper and ESA with the Juice spacecraft. And this is just the beginning. Once the potential for life in these ocean worlds becomes more widely known, the public interest will grow stronger and with it, a renewed interest in funding follow-up missions. Already, there have been proposed missions to land on Europa, return back samples from Enceladus' plume, visit Triton, and land on Pluto, just to name a few. Furthermore, in addition to these state-funded programs, very wealthy individuals have been vocal in their interest to support private missions to the ocean worlds as well. Let us first take an in-depth look at the confirmed missions from the leading space agencies.

Confirmed Missions to the Ocean Worlds

Juice

In 2012, the European Space Agency selected the Jupiter Icy Moons Explorer mission, also known as Juice, as part of its flagship L-class mission group, which will make ESA the second space agency, after NASA, to design and launch a spacecraft to the outer planet; ESA's Huygens Lander probe to Titan really just hitched a ride on Cassini. This ambitious mission will have as its primary science objective to characterize Jupiter's three icy moons—Ganymede, Callisto, and to a slightly lesser degree, Europa—and better understand their interactions with the giant planet.

The highlight of the mission will undoubtedly be Ganymede, as the final phase of the mission will see Juice orbiting the colossal moon for 9 months to better understand its magnetosphere, atmosphere, surface, and internal mass distribution. Such a continued focus will also provide insights into Ganymede's ocean that lies deep beneath the ice shelf and its habitability. Callisto and Europa will also be studied but to a lesser extent, with 12 flybys for the former and only two flybys for the latter.

Ganymede, the most prominent moon of our Solar System, is a fascinating world in its own right, see Chap. 4, and by dedicating its first ever mission to the outer-planets with a thorough study of this moon, ESA demonstrates its confidence in building complex interplanetary missions. One might wonder, though, why Ganymede was selected as the primary target of this mission instead of Europa, a far more promising moon concerning habitability and overall interest?

To address this question, we must rewind more than a decade ago, to a time when both NASA and ESA had ambitious plans. In 2008, both space agencies envisioned a grand joint mission to launch a probe to investigate one of the outer planets. As a result, two destinations were in contention. The first designation was Jupiter's icy moons under the Europa Jupiter System Mission (EJSM), also known as the EJSM/Laplace mission, whose overarching theme was the study of the emergence of habitable worlds around gas giants and understanding the interactions between Jupiter and its satellite system. The second destination was Saturn's icy moons under the Titan Saturn System Mission (TSSM), formed by the merger of two proposals: NASA's Titan Explorer and ESA's Titan and Enceladus mission (TandEM).

Ultimately, the mission to the Jovian system was deemed more promising, and TSSM was dropped. The two space agencies quickly came to the conclusion that EJSM/Laplace would be a two-spacecraft mission: NASA would focus on the "rocky" moons Io and Europa under the Jupiter Europa Orbiter (JEO) mission, while ESA would focus on the "icy" moons Ganymede and Callisto under the Jupiter Ganymede Orbiter (JGO) mission. As their names indicate, JEO was planned to orbit Europa in its final mission stage, while JGO would do the same with Ganymede. Additionally, the EJSM/Laplace mission would also investigate Jupiter, its magnetosphere, and its interaction with the satellite system by studying their interactions with the magnetosphere, gravitational coupling, and long-term tidal evolution.

It is worth noting that the Japanese space agency (JAXA) was also interested in joining the EJSM/Laplace mission and proposed the Jupiter Magnetospheric Orbiter (JMO) as well as the Jupiter and Trojan Asteroid Explorer (Trojan-JMO).

Regrettably, both missions were later canceled due to the technical challenges associated with launching them in conjunction with the JEO and the JGO.

The EJSM/Laplace mission was an ambitious one, necessitating substantial investment. Yet both agencies were optimistic that their respective governments would provide the necessary funding. In that regard, ESA's JGO mission had a significant advantage over NASA's JEO as the harsh radiation environment near Jupiter would be avoided during all JGO's mission phases. In contrast, NASA's JEO would spend most of its mission time in the harsh radiation environment closer to Jupiter, necessitating a different and more expensive design. It was estimated that JEO would be exposed to a radiation dose of 2.9 Mrad, an order of magnitude higher than JGO's 100 Krad.

Increased exposure to harsh radiation necessitated additional radiation shielding, such as aluminum and tantalum stack, to protect the sensitive scientific instruments. This shielding added more weight and, consequently, more fuel. Consequently, JEO required twice the amount of shielding (192 kg) compared to JGO (80 kg) for its 2.5-year mission, adding millions of dollars to its budget.

Furthermore, due to the high radiation levels encountered around Io and Europa, using solar panels as the primary electrical power system for JEO would be unfeasible. Instead, JEO's energy requirements would need to be provided by five nuclear batteries converting heat into electricity; multi-mission radioisotope thermoelectric generators (MMRTGs), representing the latest generation of reliable radioisotope thermoelectric generators (RTGs).[1]

The addition of the MMRTGs led to a significant rise in costs and complexity, further exacerbating the depletion of reserves of plutonium-238, the scarce fuel

---

[1] RTGs have been used on the Apollo missions to the Moon, the Viking and Curiosity missions to Mars, and all the missions to the outer planets from the Pioneers to New Horizons.

utilized to power the MMRTGs. Moreover, production of plutonium-238 had completely ceased at that time, compounding the challenge.

Furthermore, all the electronics onboard the flight systems on JEO, as well as the scientific instruments, would need to function within a high radiation environment for an extended period. This necessitated the use of custom-made components rather than off-the-shelf electronics, further escalating complexity and costs.

All the above, as well as some other cost factors, meant that JEO's estimated price tag shot up to 3.8 billion dollars; for reference, a planetary exploration flagship mission usually falls within the range of 2 billion dollars. This substantial cost, which ballooned even further to 4.7 billion dollars in a 2011 forecast, proved too hard too burdensome for the U. S. government to bear, and the mission was canceled altogether.

ESA, having received approval from European governments in the interim, was left to pursue the mission independently. In the context of the ESJM/Laplace mission, the absence of NASA's JEO resulted in a diminished scientific output regarding the study of Io, Europa, Jupiter's atmosphere, and magnetosphere.

With the cancellation of JEO, there were no alternative plans in place to study Europa, a high priority target in planetary science. An initial assessment indicated that to fully compensate for the lost scientific potential resulting from the cancellation of the Europa orbiter, JGO would need to conduct approximately fifty to a hundred flybys of the moon. However, this would necessitate sacrificing all the scientific objectives related to Ganymede, Callisto, and Jupiter. Furthermore, modifying the spacecraft to withstand the higher radiation environment would incur additional costs.

It was unanimously agreed that this wasn't a viable option. ESA determined that JGO remained pertinent to the overarching science objectives established by the ESJM/Laplace mission as the moon Ganymede, its chief target, offered ESA the opportunity to investigate a water-rich world and comprehend its interactions with the surrounding Jovian environment.

ESA concluded that the best way to maximize JGO's science return on Europa without losing focus of its foremost scientific objectives would be to add two close flybys of Europa, as well as adjust a flyby of Callisto into a close flyby of Jupiter, allowing the exploration of the Jovian atmosphere and magnetosphere at high-latitudes (30 degrees).

With the addition of these new flybys, the spacecraft would traverse through the hazardous radiation belt, resulting in an overall radiation exposure of 240 Krad. The two Europa flybys alone would contribute to a quarter of to the total mission dose. Fortunately, the updated shielding requirements for the spacecraft remained within acceptable limits for the mission. Consequently, the JGO mission was reconfigured into Juice. The key scientific objectives of Juice were redefined to focus on the study of the emergence of habitable worlds around gas giants, marking the birth of a new mission.

Juice

The Juice spacecraft was launched aboard an Ariane 5 rocket on the 14 April, 2023, from the Guiana Space Centre in French Guiana. Utilizing multiple gravity assists from the inner planets (Earth, Venus, Earth, Earth) and passing through the asteroid belt twice, Juice is scheduled to arrive at Jupiter in July 2031, after 8 years of travel.

For the first 3 years of its primary mission, it will perform a tour of the Jovian system, with close flybys of Europa, Callisto, Ganymede, and Jupiter before inserting itself into orbit around Ganymede by December 2034; it will be the first spacecraft to orbit another moon of our Solar System. There, Juice will study Ganymede for a year at ever decreasing altitudes before being disposed of on its surface at the end of 2035. Orbiting a planetary object as big as Ganymede means that the spacecraft will be eclipsed from the Sun from time to time, which will be a significant constraint on a solar-powered spacecraft. Therefore, Juice's orbits have been calculated in a way to reduce as much as possible these eclipses.

The primary scientific objectives at Ganymede include characterizing ocean layers and identifying potential subsurface water reservoirs, mapping surface topography, geology, and composition, examining physical properties of icy crusts, analyzing internal mass distribution, dynamics, and evolutionary processes, investigating the moon's tenuous atmosphere, and studying its intrinsic magnetic field and interactions with the Jovian magnetosphere.

The spacecraft hosts a suite of ten scientific instruments that will provide a full range of measurements as seen from the table below (Table 12.1). Like many large planetary missions, Juice is an international effort, with contributions from NASA and JAXA towards certain instruments (Fig. 12.1).

Juice's payload consists of ten state-of-the-art instruments plus one experiment that uses the spacecraft telecommunication system with ground-based instruments. This payload possesses the capability to fulfill all the scientific objectives of the mission, ranging from conducting direct measurements of the plasma environment to conducting remote observations of the surfaces and interiors of the three icy moons, as well as Jupiter's atmosphere.

Juice's remote-sensing package includes an imaging system (JANUS), and spectral-imaging capabilities from the ultraviolet to the sub-millimeter wavelengths (MAJIS, UVS, and SWI).

MAJIS will play a crucial role in discerning the nature of surface compounds and potentially uncovering insights into subsurface oceans. Similarly, the UVS instrument holds significance, particularly because the ultraviolet spectral regime is well-suited for studying volatiles. Many ices and relevant gases exhibit absorption patterns within this electromagnetic spectrum. We can also investigate the presence of non-ice contaminants with UV, though they can be more difficult to identify specifically.

A particularly promising avenue is the utilization of UV instruments during stellar and solar occultations, as demonstrated during the Cassini mission's study of

Table 12.1 Juice's ten scientific instruments and their science contribution.

Abbreviation	Instrument name	Description and scientific objectives
GALA	Laser Altimeter	Tidal deformation of Ganymede and morphology of moons surface features
3GM	Radio Science Experiment	Interior state of Ganymede, presence of a deep ocean and other gravity anomalies. Ganymede and Callisto surface properties. Atmospheric science at Jupiter, Ganymede, Europa ,and Callisto, and Jupiter rings.
RIME	Ice Penetrating Radar	Structure of the Ganymede, Europa, and Callisto subsurface; identify warm ice water "pockets" and structure within the ice shell; search for ice/water interface.
MAJIS	Visible-IR Hyperspectral Imaging Spectrometer	Composition of non water-ice components on Ganymede, Europa and Callisto; state & crystalinity of water ice. On Jupiter: tracking of tropospheric cloud features, characterisation of minor species, aerosol properties, hot spots and aurorae.
UVS	UltraViolet Imaging Spectrometer	Composition & dynamics of the atmospheres of Ganymede, Europa, and Callisto
JANUS	Narrow and wide Angle Camera	Local-scale geologic processes on Ganymede, Europa, and Callisto, Io Torus imaging, Jupiter cloud dynamics & structure. 'Global morphology of the Ganymede surface. Global to regional scale morphology of the Callisto and Europa surface
J-MAG	Magnetometer	Ganymede's gravity field and the extent of internal oceans. Ionosphere and upper atmosphere of Jupiter and Ganymede, Callisto, and Europa.
PEP	Particle Package	Jovian magnetosphere. Interaction between Jovian magnetosphere and Ganymede, Europa and Callisto. Exospheres and ionospheres of the moons.
SWI	Submillimetre Wave Instrument	Dynamics of Jupiter's stratosphere; vertical profiles of wind speed and temperature. Composition and structure of exospheres of Ganymede, Europa and Callisto.
RPWI	Radio and Plasma Wave Instrument	Ganymede: Exosphere and magnetosphere. Callisto & Europa: Induced magnetic field and plasma environment. Jovian magnetosphere and satellite interactions

Enceladus'plumes using UVIS. Juice's UVS may be capable of similar observations if a plume on Europa is confirmed, albeit with potential complexities.

In the past, the Hubble Space Telescope has been instrumental in studying UV emissions from all four Galilean satellites, contributing to our understanding of their atmospheres. UVS on Juice will have the capability to perform similar studies at far greater precisions.

The geophysical package consists of a laser altimeter (GALA), which will provide high-resolution maps of the moons' topography and a radar sounder (RIME) for exploring the surface and subsurface of the moons, especially in understanding the structure of the icy crusts. A radio science experiment (3GM)—using both the high gain antenna (HGA) and medium gain antenna (MGA) to probe the atmospheres of Jupiter and its satellites and to perform measurements of the gravity fields—should provide powerful insights on the distribution of the moons' interior masses.

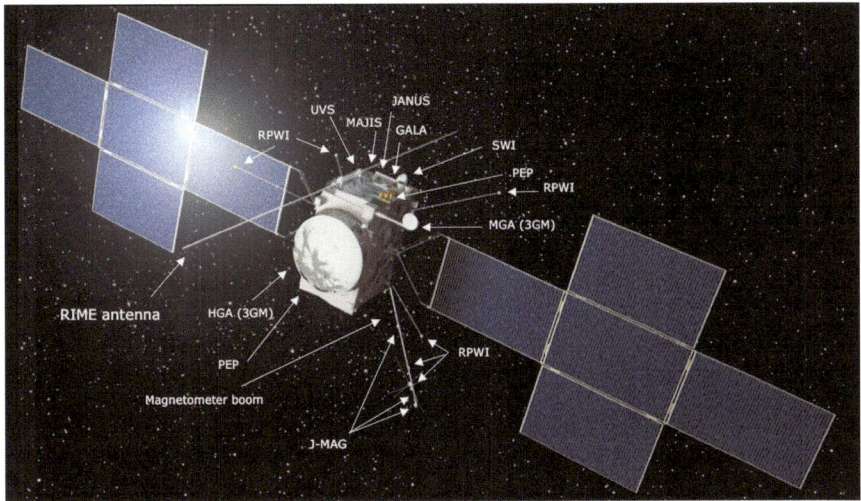

Fig. 12.1 The Juice spacecraft, equipped with 10 cutting-edge instruments, will play a crucial role in advancing our comprehension of Ganymede, Callisto, and Europa, along with Jupiter and the interactions among them. (Figure courtesy of ESA)

In addition, Juice's *in situ* package contains a powerful suite to study plasma and neutral gas environments (PEP), a magnetometer (J-MAG), a radio and plasma wave instrument (RPWI), electric field sensors, and a Langmuir probe. An experiment (PRIDE) using the ground-based Very Long Baseline Interferometry (VLBI) will support the precise determination of the spacecraft's velocity and position, with the focus on improving the ephemeris of the Jovian system.

There is no doubt that Juice is set to revolutionize our understanding of ocean worlds. Current unknowns such as the structure of Ganymede, Europa, and Callisto's subsurface as well as the composition of non-water-ice components on their surface, will be addressed. Furthermore, Ganymede will be continuously studied for a year, a first for a moon in the outer Solar System.

When it comes to Europa, the spacecraft will perform a close flyby over the northern hemisphere, followed by another over the southern hemisphere, both at very low altitudes (less than 500 km Although Juice will most likely not be able to fly through one of Europa's enigmatic plumes due to the constraints of the spacecraft's trajectories and the apparent lack of predictability in the plume's activity, PEP will be able to analyze any remnants of a hypothetical plume that might linger in Europa's exosphere, thus investigating them indirectly.

The end of the mission will be an unusual one. After more than a hundred orbits around Ganymede, the spacecraft will run out of fuel during its last orbit and make an uncontrolled crash on the surface of the moon. Ganymede is classified as a Planetary Protection Category II target, indicating that there is only a remote chance that contamination by the spacecraft could compromise future investigations. Consequently, ESA is not obligated to determine how or where the spacecraft will be deposited on the moon's surface, unlike Europa or Enceladus, which are designated as Planetary Protection Category III and IV targets, respectively.

Nevertheless, if the spacecraft is still steerable, which might be doubtful after spending many years in the harsh Jovian environment, the team might try to force Juice towards a specific location such as a flat expanse instead of a region where cracks in the icy crust are apparent.

Planetary science is, at its core, a comparative science. Juice's strength is its overarching vision of investigating and understanding the Jovian satellite system. By studying the icy moons and their interactions with Jupiter, we will gain profound insights into the three biggest oceans in our Solar System.

The Europa Clipper

As mentioned previously, following the science returned by the Galileo spacecraft in the late nineties, both the American and the European space agencies started to explore preliminary proposals for a joint mission to study the Galilean moons. NASA proposed the Jupiter Europa Orbiter (JEO) for the ESJM/Laplace mission but ultimately terminated it due to the high cost involved, exceeding 4 billion dollars.

During this period, NASA, with support from the National Science Foundation (NSF), devised a more effective approach to formulate a comprehensive strategy for exploring the Solar System: The Decadal Surveys. These surveys involve extensive collaboration within the planetary science community and aim to establish a consensus regarding the priorities for future robotic exploration missions. While NASA and the U.S. Congress, which allocates funding, are not obligated to adhere to the recommendations of the Decadal Survey, they recognize its significance and accord it serious consideration.

Both the 2003 Planetary Decadal Survey titled "New Horizons in the Solar System" and the 2011 Planetary Decadal Survey titled "Vision and Voyages" have recommended that NASA explore Europa. In fact, Europa is listed as the second-highest priority destination for a new flagship mission after Mars in the 2011 survey. Following JEO's cancelation, there was real pressure for NASA to find a way to come up with a dedicated mission to Europa, albeit within a smaller budget than what had been envisaged for JEO.

At NASA's request, new studies were undertaken in 2012 to explore the feasibility of implementing three types of missions within a strict budget of 2 billion dollars each, equivalent to the cost of a flagship mission. These mission concepts included a lander, an orbiter, and a multi-flyby spacecraft which drew inspiration from Cassini's regular flybys of Enceladus and Titan during its mission at Saturn.

The lander concept was swiftly dismissed due to the limited knowledge of Europa's surface. Despite meeting its science objectives with a 30-day orbit mission, the orbiter concept was also not chosen. The multi-flyby concept emerged as the optimal approach for fulfilling the science objectives in a cost-effective and low-risk manner, despite the spacecraft spending less than a cumulative 6 days around Europa, thus limiting its exposure to Jupiter's harmful radiations significantly.

This begged the question: how could a 6-day multi-flyby mission outcompete a comparatively longer 30-day orbiter mission? What tipped the balance in the multi-flyby mission is something we have all become familiar with in our day-to-day lives: the ability to be regularly connected to a network. Indeed, the efficiency of the multi-flyby mission lies in the spacecraft's ability to gather substantial amounts of data during each close flyby of Europa and then transmit this data back to Earth over the course of 7–10 days as it continues its orbit around Jupiter before its next pass. This allows for an almost continuous stream of data acquisition and transmission, maximizing the scientific return of the mission while minimizing operational risks.

In contrast, the orbiter spacecraft, confined to an orbit around Europa, faces challenges due to its regular passages behind the moon in relation to Earth, which obstruct transmission. Additionally, the moon's periodic passages behind Jupiter every 3.5 days further disrupt data transfer. Consequently, this configuration severely restricts the amount of data that can be sent back and thus stored, significantly impeding the orbiter's ability to transmit information effectively.

By spending a year orbiting Jupiter and making 34 targeted close flybys of Europa, as proposed in the multi-fly study, the spacecraft would have the ability to transmit three times more data compared to the 30-day period for the orbiter space-craft. It is important to note here that the data returned by the spacecraft isn't a direct multiple of time, as there are further limitations imposed by the availability of NASA's Deep Space Network antennas on Earth at any given time.

With more data to transmit, the multi-flyby mission opens the door for more data-hungry instruments such as an ice-penetrating radar or a shortwave infrared spectrometer; both essential in characterizing the moon's icy shell and surface com-position. Further studies showed that with a multiyear mission comprising 45 fly-bys, the multi-flyby spacecraft would be able to achieve most of the scientific goals set out by the initial JEO concept at half the price. And thus, born from the ashes of JEO, the multi-flyby mission was aptly rebranded as the Europa Clipper, a name derived from the fast-sailing ships of the nineteenth century, which evoke swiftness and agility. NASA had finally an answer to the community of planetary scientists urging for a return to Europa.

The Europa Clipper project was deemed highly feasible and received approval from the U.S. Congress. In 2015, it commenced its formulation phase with the aim of launching the spacecraft between 2022 and 2025. Originally, NASA had planned to launch it using the Space Launch System (SLS), which would have shortened its voyage to Jupiter. However, significant delays with the SLS program led to concerns about its reliability. As a result, the spacecraft was redesigned to be launched by the Falcon Heavy instead. At the time of writing, the Europa Clipper is scheduled to launch aboard SpaceX's Falcon Heavy on October 10, 2024.

Like ESA's Juice spacecraft, the Europa Clipper will utilize solar panels as its primary power source. This decision is driven by the fact that the spacecraft will only spend a limited amount of time within Jupiter's lethal radiation belt, enhancing the survivability of the solar panels (Fig. 12.2).

Once launched, the spacecraft will take 5.5 years to reach Jupiter using two gravity-assist maneuvers; one at Mars in 2025 and one at Earth in 2026. If all goes well, the spacecraft will insert itself in Jupiter's orbit in 2030. Europa Clipper will employ its engines as brakes to decelerate the spacecraft and synchronize its orbit with Jupiter's; this will take approximately 6 h to complete. During this deceleration process, Europa Clipper will also conduct a flyby of Jupiter's moon Ganymede.

Once in orbit around Jupiter, the Europa Clipper's prime mission is set to start in spring 2031 as it makes its first flyby of Europa. The three-year prime mission will see the spacecraft perform almost 50 flybys of Europa at altitudes ranging from 25 to 2700 km, ensuring coverage over 90% of the moon's surface.

The Clipper's scientific payload has been chosen explicitly with two objectives in mind: to characterize Europa's subsurface ocean and investigate its habitability. This includes the understanding of the processes involved in the renewal of the icy crust and the internal heat budget for the moon, as well as identifying potential plume activity with the hope of flying into one. Scientists also hope to better understand the cycling of essential elements taking place on Europa such as oxygen, hydrogen, carbon, nitrogen, phosphorous, and sulfur.

Fig. 12.2 Artist's impression of the Europa Clipper spacecraft as it approaches Europa during one of its orbits around Saturn. The Clipper is set to revolutionize our understanding of the moon. (Image courtesy of NASA)

As outlined in Table 12.2, the Europa Clipper's payload comprises a range of instruments tailored to achieve its scientific objectives.

MASPEX is set to make its debut flight aboard the Europa Clipper mission, marking the inauguration of the next generation of spectrometers in space. With

Table 12.2 The scientific payload for the Europa Clipper mission.

Abbreviation	Instrument name	Description and scientific objectives
E-THEMIS	Europa Thermal Emission Imaging System	The Europa Thermal Emission Imaging System will provide high spatial resolution, multi-spectral imaging of Europa in the mid infrared and far infrared bands to help detect active sites, such as potential vents erupting plumes of water into space.
MISE	Mapping Imaging Spectrometer for Europa	Imaging near infrared spectrometer to probe the surface composition of Europa, identifying and mapping the distributions of organics (including amino acids and tholins), salts, acid hydrates, water ice phases, and other materials.
EIS	Europa Imaging System	Visible-spectrum wide and narrow angle camera instrument that will map most of Europa at 50 meters resolution, and will provide images of selected surface areas at up to 0.5 meters resolution.
Europa - UVS	Europa Ultraviolet Spectrograph	The Europa Ultraviolet Spectrograph instrument will be able to detect small plumes and will provide valuable data about the composition and dynamics of the moon's exosphere.
REASON	Radar for Europa Assessment and Sounding: Ocean to Near-surface	Dual-frequency ice penetrating radar instrument that is designed to characterize and sound Europa's ice crust from the near-surface to the ocean, revealing the hidden structure of Europa's ice shell and potential water pockets within.
ECM	Interior Characterization of Europa using Magnetometry	The Europa Clipper Magnetometer will measure the magnetic field near Europa and in conjunction with the PIMS instrument will probe the location, depth, thickness and salinity of Europa's subsurface ocean using multi-frequency electromagnetic sounding.
PIMS	Plasma Instrument for Magnetic Sounding	The Plasma Instrument for Magnetic Sounding measures the plasma surrounding Europa to characterise the magnetic fields generated by plasma currents. These plasma currents mask the magnetic induction response of Europa's subsurface ocean. In conjunction with the ICEMAG instrument, it is key to determining Europa's ice shell thickness, ocean depth, and salinity.
MASPEX	Mass Spectrometer for Planetary Exploration	The Mass Spectrometer for Planetary Exploration will determine the composition of the surface and subsurface ocean by measuring Europa's extremely tenuous atmosphere and any surface materials ejected into space.
SUDA	SUrface Dust Mass Analyzer	Mass spectrometer that will determine the composition of the surface and subsurface ocean by measuring Europa's extremely tenuous atmosphere and any surface materials ejected into space.

over a decade of development, MASPEX has been meticulously crafted to withstand the harsh radiation environments of space and undergo the rigorous sterilization processes mandated for planetary protection protocols, including exposure to temperatures up to 300 °C.

Equipped with high-resolution capabilities, MASPEX is poised to identify small organic compounds, noble gases, and various volatile isotopes with precision. The mission's overarching goal is to detect potential biosignatures emitted by Europa's hypothetical plumes, such as distinctive concentration patterns of amino acids or fatty acids utilized in cellular membranes. Furthermore, the presence of steroids or hopanoids, which cannot be produced by abiotic processes, would serve as compelling evidence of life.

By September 2034, the Europa Clipper is planned to gently impact on Ganymede to prevent contaminating Europa. Ironically, if all goes well, there will be an overlap of a few years between the observations made by the Europa Clipper and Juice, somewhat resembling the operational synergy initially planned for the ESJM/Laplace mission. In addition, Juice will start to orbit Ganymede a few months after the Europa Clipper will have finished its observations there and impact on the moon's surface. This will allow the data from both spacecraft to complement each other, thus enhancing the science return of both missions.

There is no doubt that the Clipper mission will truly transform our understanding of Europa as an ocean world, and therefore, provide further insights into the habitability of the other ocean worlds within our Solar System. When Juice and the Europa Clipper finish their missions on the surface of Ganymede in a decade from now, the wealth of data they have gathered will provide a solid foundation for more ambitious exploratory missions to these intriguing worlds. In the upcoming section, we will delve into these missions and others in further detail.

Proposed Future Missions to the Ocean Worlds

Creating and launching planetary exploration spacecraft demands precise engineering, innovative technology, and extensive collaboration across numerous disciplines, all while navigating technical, budgetary, and scheduling challenges. Each spacecraft launched to explore our Solar System is a testament to human ingenuity, representing both a remarkable achievement and a unique endeavour. We provide here some mission proposals to the ocean worlds that might one day follow in the footsteps of Juice and Europa Clipper.

Ceres Sample Return

This mission proposes to collect samples from the Vinalia Facula evaporites in Occator crater and return them back to Earth. The spacecraft would conduct a prolonged orbit around Ceres, capturing high-resolution images for over 500 days, and akin to the OSIRIS-REx mission's approach to asteroid Bennu, it would execute a precise landing to collect samples. The return journey to Earth would span 4.7 years with the sample capsule kept at sub-zero temperatures.

The Ceres Sample Return mission would have for objective to characterize the depth and extent of potential deep brine layer(s) to determine whether liquid exists beneath Ceres today near hypothesized brine extrusion zones. It would also determine the chemistry of the waters and their potential habitability including the study of organics.

Building upon Dawn's discoveries, a return mission to Ceres would offer crucial insights into the potential activity of this dwarf planet's ocean and deepen our understanding of the complex chemistry occurring beneath its surface. Additionally, given its proximity compared to distant ocean world candidates like Europa and Enceladus, exploring Ceres would present a more accessible and cost-effective opportunity. This mission would fall under NASA's New Frontiers Program which focuses on planetary science missions with a budget cost-capped at around one billion dollars.

Europa Lander

NASA's Europa Lander was first considered in the 2010s with the aim of reaching Europa by the 2030s. Utilizing a sky crane landing system similar to the Mars Curiosity and Perseverance rovers, the lander would touch down on Europa's icy surface and endure for 20–40 days. During its stay, it would search for biosignatures from the subsurface ocean that might have been deposited on the surface.

Equipped with a robotic arm, the Europa Lander would aim to dig a 10 cm deep trench into the moon's surface, retrieving five samples of icy material from beneath the top layer. These samples would undergo analysis using advanced microscopic and spectroscopic instruments, such as a Raman spectrometer and gas chromatograph-mass spectrometer, capable of detecting organics at extremely low concentrations. The mission's objective is to identify potential microorganisms deposited on the surface by hypothetical plumes or brines. Additional instruments include stereo cameras, a seismometer, and a magnetometer to study the physical properties of the ice shell. To shield against intense radiation, all instruments would be housed within a protective vault, except for the context remote sensing instrument tasked with measuring radiation levels directly.

One potential next step involves melt-probe drilling into the icy crust to analyze pockets of liquid water situated just beneath the surface. Of course, this ambitious vision would represent a long-term goal spanning many decades to achieve.

The Europa Lander mission is still in its early design stages with some studies taking place on the engineering of the lander itself, and hasn't been formally proposed. With an anticipated cost exceeding 4 billion dollars, placing it at the upper echelon of flagship missions, it would require a substantial amount of support within the planetary science community and the U.S. congress to get considered.

A concern for the proposed lander was a study published in 2018 suggesting that photopolarimeter observations made of Europa's surface could be explained by an extremely low-density surface, formed by layers of very fine-grained particles with void space greater than about 95%. Such a surface would be less dense than snow, which would see the craft sink. If this is the case, any *in situ* investigation of the moon's surface would be delayed for decades to come. It is worth noting, though, that these photopolarimeter observations can only probe the outermost layer of the surface, which are less than a mere millimeter thick, leaving the rest of the ice below a mystery. The Europa Clipper will thankfully be able to address this point when it starts its observations of the moon.

Missions to Enceladus and Titan

Ever since jets of vapor and icy particles surging from Enceladus' South Pole were discovered almost two decades ago, numerous missions have been proposed throughout the years to study the moon in more detail, and especially sample the material being ejected into space.

Defunct mission proposals to Enceladus include the Enceladus Orbiter (EO – 2010), Enceladus Explorer (EnEx – 2012), Enceladus Life Finder (ELF – 2015), Life Investigation For Enceladus (LIFE – 2015), Enceladus Multiple Fly Mission (Moonraker – 2022), and the Explorer of Enceladus and Titan (E^{2}T – 2017).

Currently still being considered is the Enceladus Life Signatures and Habitability (ELSAH) mission proposed in 2017 for the fourth New Frontiers mission. Although the mission wasn't selected as a finalist, it received additional funding from NASA to develop cost effective techniques that limit spacecraft contamination during its construction (to eliminate false positives), with the aim of placing it in a more favorable position for the next round of New Frontiers missions which is currently delayed for the second half of the decade. ELSAH would collect minute quantities of plume substances and employ highly sensitive biosensors to scrutinize the sample for signs of life.

A more impressive mission to Enceladus being considered is the Enceladus Orbilander recommended as a flagship mission by the U.S. Decadal Survey 2023–2032. As its name indicates, the mission would act as an orbiter and lander, touching down softly on the moon's surface where it would search for evidence of life and obtain geochemical and geophysical context for the life detection experiments. The orbilander would spend over half a year orbiting the moon to sample the plume and take high-resolution images of the surface to look for a place to land. It is also planned to host a suite of life-detecting instruments such as spectrometers, microscopes and even a DNA sequencer. The spacecraft would also host radar sounders, a seismometer, and a laser altimeter.

The Orbilander would need to depart in 2038 for an arrival in the 2050s when the optimal illumination of the south polar region begins. If the Orbilander cannot be funded on time, another mission to Enceladus would also be considered: the

Enceladus Multiple Flyby (EMF). This mission would not be a flagship mission but allow the sampling of the plumes with life-detection instruments.

While all these missions to Enceladus are currently proposals, there is one confirmed mission that will visit a moon of Saturn, and that is the Dragonfly mission to Titan. Scheduled to launch in June 2028, this rotorcraft lander will explore a variety of locations on the moon and investigate the prebiotic chemistry present on the surface. Dragonfly will not investigate the subsurface ocean as such, but it will be able to indirectly gather information on what takes place below the surface with the Inertial Measurement Unit (IMU). The primary function of the IMU is for navigation and flight control.

As such, IMU constantly measures the spacecraft's acceleration, orientation, and rotation, providing real-time data to the onboard computer. This allows the spacecraft to navigate safely on Titan's surface. However, by precisely measuring Dragonfly's movement and tilts during maneuvers, scientists can analyze subtle variations in Titan's gravity. These fluctuations could suggest the existence of concealed chambers or structures beneath the surface, such as the subsurface ocean, as regions with water will exert a different gravitational force compared with ice due to their different densities.

Another mission to Titan that has been envisaged is the Titan Orbiter and Probe. Similarly to Dragonfly, this mission has for primary mission the investigation of the prebiotic environment on Titan's surface and would be able to help us understand Titan's interior structure by measuring gravity fluctuations. This mission is currently still in the proposal phase.

Missions to Uranus and Beyond

The U.S. Decadal Survey 2023–2032 has recommended as its highest-priority new flagship mission the Uranus Orbiter and Probe (UOP) with the aim to study Uranus and its system comprehensively. With a planned 4-year orbital tour of the system, UOP will provide critical insights into Uranus' origin, atmosphere, magnetosphere, and of course, its icy moons. The spacecraft will host a suite of instruments capable of characterizing in detail the surface of the moons such as imaging and infrared cameras, spectrometers as well as potentially allowing the detection of hypothetical subsurface oceans with the help of a magnetometer and by measuring the gravity fluctuations as it visits the moons. UOP would revolutionize our understanding of these icy moons and most likely confirm if Ariel is an ocean world. The best launch windows for UOP would be for 2031 and 2032, allowing a Jupiter gravity assist to shorten the journey.

The 2023–2032 Decadal Survey has also recommended a mission concept to Triton: the Triton Ocean World Surveyor (TOWS). This mission would consist of an orbiter that would perform multiple Triton flybys in a similar design to the Europa Clipper, with the aim to determine if Triton is an ocean world. TOWS would characterize the moon's interior structure as well as its surface composition and geology with the entire science payload based on existing instruments from New Horizons and the Europa Clipper such as a narrow angle camera, a multispectral imaging camera, a spectrometer, a magnetometer, a plasma for magnetic sounding, and a UV imaging spectrograph. Using launch windows in 2030 and 2031 enabling a gravity assist from Jupiter, TOWS would take 16 years to reach Neptune orbit insertion and spend 2–3 years doing 35–45 flybys of Triton. Cost estimates of this mission place it within the New Frontiers category.

Further afield, a mission concept to send an orbiter to Pluto named Perspehone has the prime science objective to determine if Pluto does indeed have a subsurface ocean. The mission would also try to unveil the internal structure of Charon and understand the evolution of the Pluto system and KBOs. It was estimated that with a launch in 2031, Persephone would require 27.6 years and a gravity assist to reach Pluto in 2058. There, it would spend 3 years to study the system with a payload of 11 scientific instruments. Given the scope of this mission, cost estimates place it within the flagship mission category.

Furthermore, alongside planning spacecraft missions to explore ocean world candidates, significant insights can be gained through observations made by Earth and space-based telescopes. The Hubble Space Telescope has already provided crucial observations of the outer Solar System, and now, the James Webb Space Telescope is doing the same. Additionally, upcoming giant telescopes like ESA's Extremely Large Telescope (39.3 meters), the Thirty Meter Telescope (30 meters), and the Giant Magellan Telescope (24 meters) will enable us to regularly observe the moons and dwarf planets of our Solar System, capturing high-resolution images and spectra. Particularly, these telescopes might aid in detecting Europa's hypothetical plumes prior to the arrival of the Europa Clipper and Juice spacecraft. This new class of giant telescopes heralds an exciting decade of observations ahead.

Ocean Worlds Program

With the multitude of proposed missions to the ocean worlds, NASA needed to establish a coordinated approach to work with scientist, engineers, budget administrators, and lawmakers. Alas, the reality of space politics, the complexities in developing and launching a space mission, and the difficulties in getting a consensus within such a broad and varied group of people has hindered the establishment of a clear roadmap in the investigation of the ocean worlds of our Solar System for the agency to follow.

There has been talk within the community on the need to establish a program similar to the Mars Exploration Program (MEP), which has been remarkably successful in ensuring the continued exploration of the Red Planet since 1993. MEP has used space probes, orbiters, landers, and rovers in response to a set of clear goals and an overarching vision. There is much to gain from a similar approach for the ocean worlds, although an additional complication would be it being a multi-target program as opposed to visiting just one object, the planet Mars in the case of MEP. While it may seem appealing to view Europa or Enceladus as isolated entities, with independent missions developed to explore them, integrating these missions into a comprehensive program would be more sensible. Insights gained from one series of missions can inform future endeavors, and technologies and instruments can be refined to meet the requirements of the entire program, as exemplified by MASPEX and MMRTG.

Furthermore, due to the long journey times required to visit ocean worlds on medium-class missions—5 years for Europa and 10 years for Enceladus or Titan—it would make sense to plan missions to prevent long gaps between each visit. Interleaving the missions will prevent the scientific community from waiting too long between new datasets as a data-starved scientific community isn't conducive for good science.

Finally, a long-term vision would prevent short-term results from negatively impacting the level of engagement and funding the program requires. For example, the disappointment of not finding any signs of life on one object would not shut off the program entirely but instead allow it to change its focus to the remaining objects. The importance of such a comprehensive strategy became apparent with Mars; after the Viking landers' disappointing results, NASA's exploration of the Red Planet stalled, leading to nearly two decades of lost momentum and neglecting a generation of Mars scientists; see Chap. 3 for additional details on the Viking mission.

With this in mind, the 2016 Congressional Commercial Justice Science and Related Agencies appropriations bill mandated NASA to establish an Ocean Worlds Exploration Program with the goal of discovering extant life in Habitable Worlds within the Solar System. To achieve this, interdisciplinary experts in Earth and Planetary science were required. In response, NASA formed the Network for Ocean Worlds (NOW) in 2019, facilitating collaboration between Earth and Planetary scientists as week as the Ocean Sciences Across the Solar System (OASS) initiative focusing on advancing research related to ocean worlds by bridging knowledge gaps and identifying opportunities for testable ideas. Given the time needed for in-depth assessments, utilizing Earth as an analogue is crucial. By uniting Earth and Planetary science communities, this initiative aims to foster comprehensive investigation and analysis, reconnecting two once-integrated communities for a unified focus.

NOW aims to achieve the following:

- To identify and stimulate novel directions of inquiry through enhanced communication within NASA's ocean worlds PI community.
- To pursue activities that both reveal and address critical knowledge gaps in ocean worlds research.
- To stimulate and facilitate new ocean worlds collaborations to undertake high-impact interdisciplinary research.
- To identify and integrate research on Earth and other ocean worlds (e.g., oceans, their interiors, and the cryosphere) to catalyze synergistic studies that identify the conditions and potential for life.
- To cultivate and augment the training of a new generation of interdisciplinary ocean worlds researchers.

Through this network, upcoming expeditions to ocean worlds could incorporate innovative technological advancements tailored specifically for the investigation and exploration of subsurface oceans.

Another promising solution to the challenges and time-consuming process of deploying costly and intricate spacecraft to explore the moons and dwarf planets of our Solar System could involve launching miniature microchips into space, akin to the Starshot initiative from the Breakthrough initiatives. Deployed in large numbers, microchips could transmit in a shorter timespan scientific data on the ocean worlds of the outer planets and distant celestial bodies like Eris, 2007 OR10, or even the far-flung Sedna.

Final Thoughts

In recent decades, Solar System exploration has yielded remarkable discoveries. While the spotlight on Mars has tempered optimism regarding its habitability, attention has shifted to numerous large moons hosting warm, salty liquid oceans. Among them, Europa and Enceladus stand out as promising candidates for harboring carbon-based life, given their subsurface oceans. Fortunately, these moons are also accessible, with Europa potentially boasting plumes and both moons featuring deep fissures within their icy crusts.

As explored in this book, other ocean worlds likely exist orbiting the large planets of the outer Solar System, with even more potentially awaiting discovery in more distant regions, such as Pluto or Eris. Indeed, subsurface oceans may be widespread throughout our Solar System, rather than exceptional occurrences. Yet, for many of them, their oceans will most likely be rich in volatile compounds such as ammonia, making it much more difficult for life to take hold.

Considering the proposed missions aimed at detecting biosignatures in ocean worlds, what if evidence emerges confirming that life has indeed taken hold in one of the subsurface bodies of water? Let us imagine a scenario where initial indicators, such as distinctive patterns in the concentrations of amino acids or fatty acids, would be discovered from subsurface material extruded onto the surface of moons like Europa or Enceladus, or within the plume of Enceladus. Further measurements of such material would be taken and after analyzing various data points and cross-checking their find, scientists would conclude that, given our understanding of carbon-based lifeforms and the geological processes presumably taking place within the planetary body, life has indeed been found.

As the initial thrill of the discovery fades, we would be faced the challenge of characterizing the newfound life, hindered by the significant technological obstacles of exploring subsurface environments with robots. Indeed, such an undertaking would likely necessitate melt probes capable of penetrating through kilometers of ice and autonomous submarines capable of sustained operation for extended periods; technologies that currently exceed our capabilities.

In the interim, we would continue living our daily lives with the awareness that life, most likely microbial, exists beneath the icy crust of a moon or dwarf planet, yet, without the ability to visit the environment where it resides, our understanding of such life would remain extremely limited. Nevertheless, would such revelation fundamentally alter humanity?

Our contemporary culture is already deeply fascinated with the notion of extraterrestrial life, with children being exposed to the idea from an early age through books and movies featuring aliens. Blockbuster films centered around extraterrestrial themes such as "E.T.", "Star Trek", and "Aliens" remain immensely popular globally, highlighting the enduring appeal of the genre. This fascination with extraterrestrial life reflects a historical tradition, as many cultures throughout history have shared stories of beings inhabiting celestial bodies.

From a scientific point of view, such a discovery would prove revolutionary. We would perform comparative studies of the biochemistry and potentially ecology of the extraterrestrial life. By assessing two biochemical systems capable of sustaining lifeforms, we would hopefully start to understand how life appeared on Earth and its universal features. For instance, if life forms on an ocean world are carbon-based, their genetic data may be stored in carbon-based molecules other than DNA/RNA. Comparisons of cell structure and organization could offer novel insights into molecular and cellular biology. Additionally, the ability to study an alien ecosystem and the processes shaping its diversity would be transformative for evolutionary biology.

Undoubtedly, the confirmation of life within an extraterrestrial subsurface ocean would raise profound societal questions about our responsibility towards this life. We would likely adhere to environmental ethics like those practiced on Earth to preserve and enhance the diversity of life. Current international rules for planetary protection focus on preventing contamination of environments for future biological explorations, rather than protecting extraterrestrial ecosystems or organisms. As a result, legislation would need to adapt considering such discoveries.

Given our expanding presence within the Solar System, we bear a responsibility to safeguard the environments and potential life forms we encounter. As we venture further outwards and potentially establish colonies on the moons of the outer Solar System in the far future, we might find ourselves custodians to this alien life, ensuring its preservation.

Our role as explorers would evolve into that of protectors, and in doing so, we might discover a new meaning to our own existence.

Appendices

Appendix A: Mimas' Mystery

Throughout the years, some have suggested that Saturn's moon Mimas might host a subsurface ocean. Let us explore this tiny moon and see if this might be possible.

With a radius of just 198 km, Mimas is notably smaller than Enceladus, making it one of the smallest known objects in our Solar System that is rounded in shape due to its own gravitation. Objects with smaller dimensions start to have potato shapes such as Hyperion (135 km radius) or are half-finished spheres such as Phoebe (106 km radius), both moons of Saturn as well.

Mimas, primarily consisting of water-ice with a small rocky core, is just massive enough to maintain a roughly spherical shape, albeit slightly elongated due to Saturn's gravitational pull. Its dimensions reveal a slight egg-shaped ovoid form, with the diameter facing Saturn stretching 9% longer than the diameter perpendicular to its orbit, measuring at 209 × 196 × 191 km. The moon bears the scars of numerous impact craters, with one side notably marked by a massive crater named Herschel, in honor of the moon's discoverer. This crater measures 139 km across and 5–7 km deep, with its walls towering an additional 5 km high, making it one of the most prominent features on Mimas and one of the largest relative to its parent body in the entire Solar System.

Prior to the Space Age, Mimas remained a distant and mysterious speck of light. However, this changed on July 31, 1979, when Pioneer 11 conducted a flyby of the moon, capturing an image as Mimas transited in front of Saturn. Unfortunately, the image was of poor quality. It wasn't until a year later, with the arrival of Voyager 1 and its more advanced imaging system, that planetary scientists were able to reveal a clearer view of Mimas.

Voyager 1's observations enabled scientists to determine Mimas'density at 1.15 g/cm³, a value close to that of water. This suggested that the moon was predominantly composed of water-ice, distinguishing it from other icy moons of Saturn

B. Henin, *Exploring the Ocean Worlds of Our Solar System*, Astronomers' Universe, https://doi.org/10.1007/978-3-031-62953-2

known to have significant rocky cores, such as Enceladus, Tethys, Rhea, Dione, and Iapetus. Mimas was categorized as a unique type of moon, often referred to as a "giant snowball."

Despite their compositional differences, scientists felt compelled to compare Mimas with Enceladus, given their relatively similar sizes and location. In addition, both moons are in resonance with another moon—Mimas with nearby Tethys and Enceladus with Dione—and both display eccentricities in their orbits, with Mimas' eccentricity at 0.0196, four times bigger than Enceladus' at 0.0047. This was a surprising find. According to a paper published in 1983, given the parameters above, "Mimas should currently be tidally heated at a rate at least twice that of Enceladus if Mimas' rigidity is like that of rock, and as much as 30 times if its rigidity is the same as Enceladus."

Nevertheless, Mimas shows no evidence of recent tectonic activity. Indeed, Voyager 1 revealed the moon to be solidly frozen at a temperature of 64 K (−209 °C). This became known as the 'Mimas paradox' or 'Mimas test.' As much as Enceladus' surprisingly active geology required explanation, Mimas' inactivity was as compelling. Any theoretical models put forward to account for Enceladus' characteristics also had to do the same for Mimas and vice versa. This paradox proved to be frustratingly tricky for planetary scientists and would only be solved decades later, thanks to new data from the Cassini spacecraft.

Another surprise from the Voyager flyby was Hershel Crater. It is not unique within the middle-size icy satellites, as Tethys, Dione, Rhea, and Iapetus also host several large impact basins whose diameters are a substantial fraction of the satellite's diameter. (Enceladus is the odd one out here due to its younger surface.) However, Hershel is the most remarkable. At one-third of the moon's diameter, it is the largest in relation to the size of the moon. In some parts, it is 12 km deep and hosts a central peak which rises to 8 km in height. As craters go, this one is extremely deep.

What is more, large troughs similar to shock waves 10 km wide were found across the moon's surface, suggesting global-scale fractures created by the Hershel impact event. If the impactor had been a little bit bigger or come at a faster speed, it might have broken up the moon, which most likely would have ended up as one of Saturn's ring.

With such upheaval and no signs of past or present geological activity, how could this small moon have been considered by some to host a subsurface ocean?

Observations from the Cassini orbiter during its 13 years within the Saturnian system returned never seen before images. For a start, it was found that due to its position, and despite its low mass, tiny Mimas was responsible for the Cassini division, a 4800 km wide gap between Saturn's A and B rings. Actually, Mimas is locked in resonances with many objects or features within the Saturnian system, such as the Huygens gap, the G-ring, objects lying between the C and B ring, nearby moons Dione and Enceladus as well as with the larger moon Tethys (2:1 resonance) and the tiny moon Pandora located in the outer F Ring (2:3 resonance).

The precise measurements of all these complex interactions allowed scientists to finally resolve the Mimas paradox. Indeed, the moon's resonance with Tethys, which was initially thought to add tidal heating to Mimas (in a similar way Enceladus is in resonance with Dione), is a different type of resonance and isn't responsible for the tiny moon's eccentricity.

Instead, the resonance both moons share is related to the inclination in their orbits, as these are tilted with respect to the orbital planes of Saturn's satellite system. As they orbit Saturn, Mimas orbits twice for each orbit of Tethys, they meet up not at the closest point of Mimas' orbit (periapsis) but instead at multiple locations throughout their orbits. As a result, Tethys doesn't pull Mimas into another orbit and is not responsible for the moon's eccentricity. In fact, when we take into account all the interactions Mimas has with the objects orbiting Saturn, we find no source for the moon's eccentricity, suggesting that it is most likely a leftover process, a fossil from earlier times, when the moons were in different orbits. (Saturn's spin pushes the moons further away with time, thus altering their orbits.) Recent simulations have shown that around 2 billion years ago, Mimas might have gone through a 2:3 resonance with Enceladus, giving it a much higher eccentricity at the time that has decayed ever since.

Furthermore, the high-resolution images from the Cassini orbiter have also revealed that no surface was left intact by the intense bombardment Mimas experienced throughout the ages. With no internal processes to erode or erase them, the frozen surface has preserved the craters for billions of years.

However, by carefully studying the surface craters, it was found that the south pole region hosts craters half the average size ranging from 20 km in diameter or less, hinting at possible resurfacing processes at some point in the moon's life. (Coincidentally, Enceladus' most active region is also located at the south pole.) In support of such interpretation, the depth of the Herschel Crater and soft features observed on its rims indicate that it formed as a flexible, slushy surface, hinting that the moon's surface and interior might have been much warmer in the past, most likely due to greater orbital eccentricity.

An increased eccentricity must have pumped heat into the moon, softening it and giving it its round shape. Some scientists have therefore speculated that it could have potentially been warm enough to allow its small icy interior to melt and form a small subsurface ocean.

Although Mimas' paltry size and lack of meaningful rocky core meant that it probably couldn't retain a subsurface ocean for long periods of time, this idea raised eyebrows among Cassini mission researchers, as a perplexing pattern in the moon's motion was discovered; Mimas was wobbling. Referred to as libration, this perceived oscillation motion might reveal what lies inside the moon. The properties of a raw and a hard-boiled egg are often used to explain this concept. If you place both eggs on the table and spin them, you will notice that the hard-boiled egg can spin evenly and at a fast pace, while the raw egg will be slower and spin unevenly as the white and yolk slosh around inside.

By using images returned from Cassini, initial studies published in 2014 found that Mimas wobbles twice as much as predicted if it had a typical solid interior. The study concluded that this could only be explained by two possibilities: either Mimas contains a frozen interior with a non-spherical elongated core in the shape of a rugby ball or an American football, or it hosts a subsurface ocean (like the sloshed liquid inside a raw egg). Both possibilities have problems, though. Hosting a non-spherical core is not what would be expected from a planetary object billions of years old, as central cores relax into a spherical shape with time. On the other hand, the presumed existence of a subsurface ocean was puzzling as well, since the moon hasn't shown any substantial geological activity on its surface for billions of years, and its tiny size should make it impossible to retain heat for a significant period. The study showed that if a subsurface ocean was indeed present, it should lie within 24–31 km below the surface and be global. Although most planetary scientists remained skeptical about this study, some were hopeful that Mimas could be an ocean world candidate.

Alas, all this changed when a new paper published in February 2017 put a blow to the subsurface ocean theory. In this new study, it was calculated that the stresses on Mimas' icy crust induced by a subsurface ocean were much too strong and would have produced large surface fractures within the crust over time. Since such fractures have not been observed on the moon's surface, it is highly unlikely that Mimas hosted a subsurface ocean. Instead, the moon's libration was best explained by it possessing a small silicate core that initially started as a sphere but was later pushed askew by a strong impact such as the one creating Herschel Crater, giving the moon such an asymmetric angular moment.

It seems therefore that despite Mimas experiencing past tidal heating, a subsurface ocean has most likely never been formed. Future missions to study Mimas will hopefully provide conclusive evidence that its libration is indeed induced by an ovoid core.

Appendix B: Relic Surface Oceans

Three Water Worlds

Although the main coverage in this book is of the subsurface oceans in our Solar System, there is a sense of perspective to be gained by reviewing planets that had oceans on their surfaces in their past or present.

As explained in chapter two, Earth, Mars, and Venus were bombarded by ice-rich bodies after their formation, allowing them to amass a substantial amount of water on their surfaces. Now, let us imagine these planets as water worlds with deep blue oceans present on their surfaces and ringed by billowing white clouds drifting high up in their thick atmospheres. On Venus, two continents rise above the water: Ishtar Terra and Aphrodite Terra. Although the latter is the largest, the former hosts the highest peak, towering 11 km above sea level. On Mars, the northern hemisphere is entirely covered by water, while the southern hemisphere is a giant continent by itself containing a vast inland sea residing inside the most prominent crater on the planet, Hellas Basin. And finally, the last of these water worlds holds a vast ocean upon which a supercontinent lies, waiting to be partitioned in the world we know today.

Some scientists have proposed that roughly 5 hundred million years after their formation, Venus, Mars, and Earth enjoyed similar if not identical environments for tens of millions of years, including vast oceans, surface temperatures above freezing, and an atmosphere where clouds cover the globe.

Interestingly, as we have seen in chapter three, there are tantalizing clues that life might have already appeared on Earth during this period. Could life have started on Venus and Mars as well? It is an intriguing thought as cross-seeding might have occurred between the three planets. We might be Martians or Venusians.

But that was then. Nowadays oceans of water are not what one immediately pictures when thinking of our neighboring planets. Venus, for instance, is shrouded in a dense atmosphere, experiencing average surface temperatures of 462 °C and atmospheric pressures 92 times greater than Earth's. This extreme environment resembles a planet-sized pressure cooker, rendering the surface one of the driest locations in the Solar System.

Mars, on the other hand, is the opposite. Lacking a dense atmosphere, low pressures inhibit liquid water on the surface, as it will sublime directly into water vapor. Therefore, most of the planet's water is trapped in polar caps or underground ice. Luckily for us, Earth had the right conditions to sustain its surface oceans for billions of years. So, what happened to the primordial oceans of Venus and Mars? Let us review them in detail.

Blue Mars

Mars is the most well-understood planet in our Solar System after our own. Although this isn't saying much, since there are still large gaps in our knowledge, it does illustrate how the second smallest planet in our Solar System has fascinated us ever since we looked up at the night sky. The figures speak for themselves; at the time of writing Mars had been visited by over 55 spacecraft (taking into account all various flybys and gravity assists), making it the most visited object in our Solar System closely followed by Venus with over 40 missions. Out of those 55 Mars missions launched since the 1960s, only half succeeded in achieving their primary science goal (see Figs. 3.1 and 3.2 in Chap. 3).

The USSR and subsequently Russia hold the unenviable title for the most failed missions, with a total of twenty out of twenty-two. Ironically, the only two Russian spacecraft that did manage to orbit Mars in the early 1970s did so while an unexpected dust storm raged on the entirety of the planet, limiting the scope of their science mission.

Out of the successful missions, though, we've had orbiters, flybys, landers, and car-sized rovers. NASA holds the lion's share by launching most of these missions, ranging from its first flyby that lasted 2 days (July 14–15, 1965) to its longest-serving planetary robot, Opportunity, which was active for more than 14 years on the surface (as opposed to its original planned mission duration of only 3 months).

What have all these robotic emissaries taught us about Mars' past? We now know that it was wet and remained so for a period. The amount of liquid water and the time this water stayed on the surface is still open for debate among planetary scientists, yet it seems that Mars might have had an ocean and maybe two. Let us review the evidence.

To start off with, due to the planet's small size, its interior cooled off rather quickly, which brought to a halt tectonic and volcanic activity. As a consequence, the planet's crust solidified early on in its history, contrary to our planet, which regularly resurfaces the crust every few hundred million years. Consequently, original features that were erased a long time ago on Earth's surface remain relatively unchanged on Mars, allowing us to travel back in time and analyze rocks and geological formations that are billions of years old, a rarity on Earth. Given this, if ancient oceans did exist on Mars's surface, one would expect that to see visible evidence such as shorelines, deltas, rivers, and channels which would be feeding into these oceans.

Already, when the NASA Viking orbiters sent back detailed images of the planet's surface in the 1970s, some researchers thought they had detected ancient shorelines along the boundary between the northern and southern hemispheres. Not everyone was convinced, though, as the evidence was weak at best and subject to interpretation. Images returned from later orbiters weren't conclusive either despite unprecedented imaging capability. However, new lines of evidence uncovered in the last decades seem to support the ocean hypothesis.

One such piece of evidence includes numerous regions within the northern hemisphere where images show remnants of deep channels carved by rain as well as the existence of lakes. Such features could be explained if a large body of water was present for a significant amount of time to bring about the conditions required for cloud formation and rainfall. In addition, many ancient deltas were observed at an altitude where the shoreline was thought to be situated by the ocean hypothesis. These deltas, characteristic of a river entering slow-moving or standing water, suggest that this theoretical yet unseen shoreline remained stable for a long period.

In 2012, the European Space Agency published results collected by the Mars Express orbiter revealing a subsurface blanket of low-density material around the northern polar cap. Contrary to the southern hemisphere, which is comprised of hardened volcanic flows, the presence of low-density material in the northern hemisphere, potentially rocky material mixed with ice, suggests sedimentary material, tens of meters thick. This supports the idea that material was deposited on an ocean floor due to standing water.

What is more, in 2015, after 6 years of atmospheric observations, scientists found a high ratio of deuterium in the planet's atmosphere indicative that ancient Mars contained much higher water levels than it does today. As you might recall from chapter two, deuterium is the hydrogen isotope that forms heavy water molecules. In the past, as these molecules of water evaporated from the surface, they encountered lethal solar radiation high up in the atmosphere and got split in the process. The oxygen dissipated into space while the hydrogen isotope accumulated in the atmosphere, acting like a marker. Measuring its concentration in the current atmosphere not only reveals that water molecules were present in the planet's past but also allows us to extrapolate how much quantity there was. Indeed, since water on Earth and Mars started off with the same D/H ratio, we can measure the difference and calculate how much 'light water' was lost. And the figure is telling.

The concentration of deuterium in Mars' atmosphere is about eight times as much as on Earth. This points to a significant loss of water over time, with some models suggesting that Mars had enough water to cover the planet to a depth of 137 meters. All this water must have accumulated in an ocean at the lowest point on the planet, its northern hemisphere. The reason for its disappearance is one of the areas that is still being researched, but it is commonly agreed that Mars' lack of a protective magnetic field prevented its nascent atmosphere from withstanding the continuous blows from the solar wind, stripping it away during millions of years. This, in turn, reduced the atmospheric pressure that led to the slow but inevitable evaporation of the surface water as well as a substantial drop in the temperature, forcing any remaining freezing water to stick to the ground.

Finally, recent discoveries have also shed new light on the paradox of the perplexing lack of clearly defined shorelines. Thanks to the resolution power of NASA's HiRISE; a powerful telescope orbiting the Red Planet, scientists discovered unique surface formations dotted along the boundary between the northern and southern hemispheres. On Earth, these features are mounds of deposited sediments and are called thumbprint terrain. It was previously thought that they were the result of glaciers or mud moving downhill from volcanoes, but it has now been shown to be a leftover feature of one or multiple tsunamis hitting the shorelines. Finding these thumbprint terrains on Mars has led some scientists to suggest that over 3 billion years ago, a giant asteroid hit the planet in what was once the northern hemisphere ocean. An asteroid impact could create multiple tsunamis that would have plowed the coastline of the ancient ocean and buried its shorelines with large deposits. In support of such claim, it has been suggested that the impact site for such an event was Lomonosov Crater, a 120 km wide bowl in the northern hemisphere. Such a hypothesis not only provides further evidence for the existence of an ocean, as tsunamis require vast amounts of water to be created, but it would also finally explain why the ancient shorelines haven't been found.

More scientific data will be collected by future Martian missions, allowing scientists to characterize this possible ocean with much greater certainty and detail. Many unanswered questions persist, including inquiries into its temperature, composition, evolutionary changes, and interactions with the atmosphere.

This very brief outline of the likelihood that an ancient surface ocean was present on the fourth planet from our Sun doesn't do justice to this fascinating subject. Many intriguing points could be explored such as the possibility of finding substantial amounts of water-ice hidden under a thin layer of dust in the northern hemisphere, the possible leftover of the frozen ocean. In August 2024, scientists revealed that seismic data gathered between 2018 and 2022 by NASA's InSight lander enabled the detection of large volumes of liquid water reservoirs located at depths of approximately 10 to 20 km within the Martian crust. According to this new study, igneous rocks saturated with liquid water best explain the existing InSight data. This suggests there could be much more water and more importantly, liquid water, trapped within Mars than we previously expected, raising the likelihood of Martian life.

Regarding an ancient surface ocean, recent models suggest that it could have been sustained even if the average temperature on the planet was below zero with ice sheets and glaciers present. Mars could have been cold but wet. On the other hand, some scientists are still not convinced of the ocean hypothesis, as some models have a hard time sustaining an atmosphere capable of supporting a surface ocean for an extended period.

Even though our goal in this book is not to cover this topic in great depth, this brief overview on Mars showcases how a systematic and comprehensive exploration program of a planetary body such as MEP can provide multiple lines of evidence that complement each other and build a better understanding of the planet as a whole. It also highlights the vulnerability of surface oceans, which can be disrupted or even lost, if not by catastrophic events, then by the slow disappearance of a protective atmosphere. Let us now visit the second planet to the sun, Venus, as it also has a story to tell, one that demonstrates the inherent difficulties of space exploration.

Blue Venus

Contemplating the existence of a liquid water ocean on Venus may appear absurd due to its current extreme temperatures, which preclude the presence of liquid water. Nonetheless, scientists are entertaining the notion that Venus experienced a wetter epoch lasting hundreds of millions or even billions of years in the past. Substantiating this hypothesis encounters two unavoidable challenges though.

Firstly, due to the harsh conditions present on the surface, no spacecraft, lander, or rover will be capable of investigating the surface of Venus in a similar way that we have methodically explored Mars throughout the last decades. And while the engineering challenge is an interesting one, the high cost of building and sending a robotic probe and lander capable of surviving the Venusian surface for extended periods of time would bring sleepless nights to any financial planners. There will never be a 'Venusian Perseverance rover' busy exploring the surface for numerous years.

Regardless, such a mission is not required, as—and this is the second point—Venus has a very dynamic geology and experiences regular extensive volcanic activity resurfacing parts of the crust, if not all. In complete contradiction with Mars, which has its past exposed in the open for anyone curious enough to investigate, Venus has erased many of the surface evidence of its distant past, leaving little hope for researchers eager to study such features.

So, if we can't see shorelines, deltas, channels, and sedimentary rocks, what makes scientists confident in their assertion that Venus was once a blue planet? The case for past Venusian oceans derives from our understanding of the formation of our Solar System. In effect, the way we appreciate the planet today has benefited from the comprehensive robotic exploration of the Solar System carried out by the major space agencies. From studying asteroids, comets, and the inner planets, and establishing theories on how these bodies were formed, we have learned to uncover Venus' past.

In chapter two, we explored the idea that most inner planets were pounded by water-rich asteroids and sometimes comets, and both Mars and Earth held vast amounts of water. Venus was no exception. The fact that it resides a bit closer to our Sun than Earth or Mars doesn't change the fact that it was also composed of the same stuff. Therefore, Venus also must have started with deep oceans in the early part of its history. Venus was initially a blue planet.

Luckily, in those early years, our Sun was dimmer, according to the standard model, roughly 40% less bright than it is today. Therefore, Venus received less heat from solar radiation, and models show that it could have sustained oceans on its surface for a very long time. For how long? We don't know. Maybe for a few hundreds of millions of years to a billion years. Once the Sun started to increase its energy output, more sunlight hit Venus' thick atmosphere, trapping an increasing amount of heat, and warming it up.

This started an evaporation process that sent huge amounts of water vapor into the atmosphere. Similar to Mars, Venus lacks a magnetic field, and thus, a means to protect itself from the constant radiation of our Sun. For example, ultraviolet radiation from our Sun collided with the water molecules high up in the atmosphere and broke them apart, resulting in oxygen molecules leaking out into space. Little by little, Venusian oceans evaporated into the atmosphere with some parts blown away into space. Luckily for us, this process has left a trace in Venus' atmosphere, and we have been able to measure the deuterium ratio, as we have done so on Mars. Scientists have found a high D/H ratio within Venus'atmosphere today, a clear indicator that the planet had a much wetter past capable of supporting oceans. Further exploration of this fascinating planet is required if we want to unveil Venus'relic ocean.

Once more, the subject of past oceans on Venus proves intriguing, deserving of deeper exploration beyond the limited coverage provided here. Despite lingering uncertainties, it underscores the fragility of surface water bodies on planetary worlds.

Conversion Tables

Temperature scales		
Kelvin (K)	Celsius (°C)	Fahrenheit (°F)
0	−273	−460
25	−248	−414
35	−238	−397
50	−223	−370
75	−198	−325
100	−173	−280
125	−148	−235
150	−123	−190
173	−100	−148
233	−40	−40
255	−18	0
273	0	32
293	20	68
310	37	99
373	100	212
423	150	302
473	200	392
773	500	932
1273	1000	1832
2273	2000	3632

Distance scales	
Kilometers	Miles
1	0.6
50	31.1

(continued)

© The Editor(s) (if applicable) and The Author(s), under exclusive license to Springer Nature Switzerland AG 2024
B. Henin, *Exploring the Ocean Worlds of Our Solar System*, Astronomers' Universe, https://doi.org/10.1007/978-3-031-62953-2

Distance scales

Kilometers	Miles
100	62.1
150	93.2
200	124.3
250	155.3
300	186.4
350	217.5
400	248.5
450	279.6
500	310.7
650	403.9
700	435.0
750	466.0
800	497.1
850	528.2
900	559.2
950	590.3
1000	621.4
1100	683.5
1200	745.6
1300	807.8
1400	869.9
1500	932.1
2000	1242.7
3000	1864.1
4000	2485.5
5000	3106.9
10000	6213.7

Astronomical Unit

AU	Kilometers
1	149,597,871
2	299,195,742
3	448,793,613
4	598,391,484
5	747,989,355
10	1,495,978,710
15	2,243,968,065
20	2,991,957,420
30	4,487,936,130
40	5,983,914,840
50	7,479,893,550
100	14,959,787,100

Glossary

Albedo (meaning "whiteness") The measure of the solar radiation reflected back from a planetary object.

Archaea One of the three great domains in life (bacteria and eukaryotes are the other two), these simple life-forms lack a nucleus to store their DNA. Archaeans include inhabitants of some of the most extreme environments on the planet and may be the only organisms that can live in extreme habitats such as thermal vents.

Astrobiology The study of the origin, evolution, distribution, and future of life in the universe. It lies at the interface between biological sciences and planetary sciences.

Bacteria One of the three great domains in life (archaea and eukaryotes are the other two), these simple life-forms lack a nucleus to store their DNA.

Biosignature Any phenomenon produced by life.

Core The planetary core consists of the innermost layer(s) of a planetary object and may be composed of solid or liquid matter.

Crust the outermost solid shell of a planetary object. It is usually distinguished from the underlying mantle by its chemical makeup; however, in the case of icy satellites or dwarf planets, it may be recognized based on its phase (solid crust vs. liquid mantle).

Differentiation The transformation of a homogenous body into a heterogeneous body. If a planetary body is large enough it will develop a core, mantle, and crust, each of which may be further subdivided. Each layer of Earth has its own set of subdivisions, for example upper, middle, and lower crust.

Eccentricity The orbital eccentricity of an astronomical object is a parameter that determines the amount by which its orbit around another body deviates from a perfect circle. A value of 0 is a circular orbit, values between 0 and 1 form an elliptical orbit, 1 is a parabolic escape orbit, and greater than 1 is a hyperbola.

Extremophile Any organism (particularly microorganisms) that inhabit extremes of chemical or physical conditions.

B. Henin, *Exploring the Ocean Worlds of Our Solar System*, Astronomers' Universe, https://doi.org/10.1007/978-3-031-62953-2

Frost line (Snow line or ice line) location in our Solar System where it is cold enough for volatile compounds such as water, ammonia, methane, carbon dioxide, and carbon monoxide to condense into solid ice grains.

Habitability The potential of a planetary body to have habitable environments hospitable to life, or its ability to generate life endogenously.

HP ice or high-pressure ices As water-ice (1 h at P = 1 atm) is compressed at low temperatures, it undergoes a series of phase transitions between different molecular structures.

Hydrothermal vents Sources of hot, mineral-rich waters located in fractures on deep-ocean submarine ridges. One of the candidates for the emergence of life on Earth.

Late heavy bombardment (LHB) A period from around 4 to 3.8 billion years ago when intense comet and asteroid bombardment occurred.

Mantle The layer between the crust and the outer core. It is often divided into layers of different composition.

Ocean world A planetary object that hosts a subsurface ocean of liquid water (and other non-water components).

Organic chemistry The study of the carbon-based structures, properties, and reactions of matter in its various forms.

Panspermia The theory that life on Earth originated from microorganisms or chemical precursors of life present in outer space and able to initiate life on reaching a suitable environment.

Peroxides Any class of compounds in which two oxygen atoms are linked together by a single covalent bond.

Photochemistry The study of chemical processes that occur because of the absorption of light.

Planetary body A term used to describe planets, satellites, and asteroids.

Planetary protection The prevention of the contamination of other planetary bodies or the contamination of Earth with extraterrestrial organisms.

Serpentinization An exothermic chemical reaction between rocks (rich in magnesium and iron) and water, giving rise to strongly alkaline fluids saturated in hydrogen gas.

Spectra (Plural of *spectrum*) The full range of all frequencies of electromagnetic radiation.

Subsurface ocean A large body of liquid water lying underneath an icy crust or mantle of a planetary object (mainly in icy satellites or dwarf planets).

Tidal heating Orbital energy dissipated as heat in either a surface ocean or the interior of a planet or satellite.

TNO (Trans-Neptunian Object) Any planetary body in the Solar System that orbits the Sun at a greater average distance (semi-major axis) than Neptune, 30 astronomical units (AU). This includes the Kuiper Belt and the Scattered Disc.

Tholin Brownish-red substances made of complex organic compounds.

Volatiles Elements or compounds that melt or boil at relatively low temperatures. Examples include hydrogen, helium, methane, and water.

Further Reading

Books

Alien Oceans: The Search for Life in the Depths of Space, by Kevin Hand (Princeton University Press, 2020)

Alien Seas: Oceans in Space, by Rosaly Lopes & Michael Carroll (Springer, 2013)

Alien Volcanoes by Rosaly Lopes & Michael Carroll (Johns Hopkins University Press, 2008).

Analog Experiments for the Identification of Trace Biosignatures in Ice Grains from Extraterrestrial Ocean Worlds. Klenner, F. et al. Astrobiology 20, 179–189 (2020a).

An Introduction to the Solar System (3rd Edition) by David A. Rothery, Neil McBride & Iain Gilmour (Cambridge University Press, 2011).

An Introduction to Astrobiology (3rd Edition) by David A. Rothery, Iain Gilmour & Mark A. Sephton (Cambridge University Press, 2018).

Asteroids: Relics of Ancient Time by Michael K. Shepard (Cambridge University Press, 2015)

Astrobiology: Understanding Life in the Universe by Charles S. Cockell (Wiley Blackwell, 2015).

Cassini-Huygens (NASA/ESA/Asi) – Owners Workshop Manual by Ralph Lorenz (J H Haynes & Co Ltd, 2017).

Discriminating Abiotic and Biotic Fingerprints of Amino Acids and Fatty Acids in Ice Grains Relevant to Ocean Worlds. Klenner, F. et al. Astrobiology 20, 1168–1184 (2020b).

Enceladus and the Icy Moons of Saturn (Space Science Series) by Paul Schenk, Roger Clark, Carly Howett, Anne Verbiscer, Hunter Waite (University of Arizona Press, 2018).

Europa (Space Science Series) by Robert T. Pappalardo, William B. McKinnon, Krishnan Khurana (University of Arizona Press, 2008).

Foundations of Astronomy, Enhanced (13th Edition) by Dana Backman & Michael Seeds (Brooks Cole, 2015).

Imaging the Solar System by Bernard Henin. Springer Publishing (2022)

Jupiter: The Planet, Satellites and Magnetosphere by Fran Bagenal (Cambridge Planetary Science, 2007).

Ocean Worlds: The Story of Seas on Earth and Other Planets by Jan Zalasiewicz & Mark Williams (Oxford University Press, 2018).

Physics and Chemistry of the Solar System (2nd Edition) by John S. Lewis (Academic Press, 2012).

Planetary Geology: An Introduction (2nd Revised Edition) by Andrew Dominic Fortes & Claudio Vita-Finzi (Dunedin Academic Press, 2013).

Planetary Sciences (Updated 2nd Edition) by Imke de Pater & Jack Lissauer (Cambridge University Press, 2015).

Planets and Moons: Treatise on Geophysics by Tilman Spohn (Elsevier Science, 2009).

Mass Spectrometric Fingerprints of Archaea and Bacteria for Life Detection on Icy Moons. Salter, T., H. Wait, and M.A Sephton (2020) AGU Fall Meeting 2020, Abstract # P001-03

Moon Hunters: NASA's Remarkable Expeditions to the Ends of the Solar System by Jeffrey Kluger (Simon & Schuster, 2001).

NASA'S Voyager Missions: Exploring the Outer Solar System and Beyond (2nd Edition) by Ben Evans (Springer, 2008).

Neptune and Triton by Dale P. Cruikshank, Mildred Shapley Matthews & Dale P. Cruikshank, A. M. Schumann (University of Arizona Press, 1995).

Robotic Exploration of the Solar System: Part I: The Golden Age 1957–1982 by Paolo Ulivi & David M. Harland (Springer, 2007).

Robotic Exploration of the Solar System: Part 2: Hiatus and Renewal, 1983–1996 by Paolo Ulivi & David M. Harland (Springer, 2008).

Robotic Exploration of the Solar System: Part 3: Wows and Woes, 1997–2003 by Paolo Ulivi & David M. Harland (Springer, 2012).

Robotic Exploration of the Solar System: Part 4: The Modern Era 2004–2013 by Paolo Ulivi & David M. Harland (Springer, 2014).

The Cambridge Guide to the Solar System by Kenneth R. Lang (Cambridge University Press, 2011).

The Ringed Planet: Cassini's Voyage of Discovery at Saturn by Joshua Colwell (Morgan & Claypool, 2017).

The Rivers of Mars: Searching for the Cosmic Origins of Life by Piers Bizony (Aurum Press Ltd, 1997).

The Science of Solar System Ices by Murthy S. Gudipati & Julie Castillo-Rogez (Springer, 2012).

The Vital Question: Energy, Evolution, and the Origins of Complex Life by Nick Lane (W. W. Norton & Company, 2016).

NASA Voyager 1 & 2 Owners' Workshop Manual (Including *Pioneer 10 & 11*) by Christopher Riley (J. H. Haynes & Co Ltd, 2015).

Scientific Papers and Space Agency Reports

"Abiotic and Biotic Formation of Amino Acids in the Enceladus Ocean: Speculation on the annual biomass production and cell concentrations in Enceladus' ambient ocean based on the inferred internal hydrothermal activity" by Elliot Steel, Alfonso Davila & Christopher McKay. *ASTROBIOLOGY* Volume 17, Number 9, 2017.

"Can Life Begin on Enceladus? A Perspective from Hydrothermal Chemistry: The case for the origins of life in surface hydrothermal fields as opposed to deep-sea vents" by David Deamer & Bruce Damer. *ASTROBIOLOGY* Volume 17, Number 9, 2017.

"Experimentally Testing Hydrothermal Vent Origin of Life on Enceladus and Other Icy/Ocean Worlds" by Laura M. Barge & Lauren M. White. *ASTROBIOLOGY,* Volume 17, Number 9, 2017. This paper reviews the laboratory strategies and methods that can be utilized to simulate the origin of life in hydrothermal vent systems on icy/ocean worlds.

"Explorer of Enceladus and Titan (E^2T): Investigating ocean worlds' evolution and habitability in the solar system" by Giuseppe Mitri et al. *Planetary and Space Science* (2017) 1–18. In depth review of the science case for the exploration of Enceladus and Titan with an M-class ESA mission.

"Follow the Plume: The Habitability of Enceladus" by Christopher McKay, Ariel Anbar, Carolyn Porco, and Peter Tsou. *ASTROBIOLOGY,* Volume 14, Number 4, 2014. A study focusing on the search for biomolecular evidence of life in the organic-rich plume of Enceladus.

"Heat Transport in the High-Pressure Ice Mantle of Large Icy Moons" by G. Choblet, G. Tobie, C. Sotin, K. Kalousová, & O. Grasset. *Icarus* 285 (2017) 252–262. Paper on the properties of high-pressure ices in contact with a rocky core, and the emergence of hot convective plumes transporting minerals to the above ocean.

Juice definition study report (Red Book). (ESA website) Everything you ever wanted to know about JUICE published by ESA in November 2016.

"Ocean Worlds Exploration: A case for the exploration of the ocean worlds of our Solar System" by Jonathan I. Lunine. *Acta Astronautica*, November 2016. This paper was instrumental in shaping the structure of this book.

"Ocean Worlds White Paper: Ocean Sciences Across the Solar System" NASA. https://ocean-worlds.space/whitepaper/

"Pluto's ocean is capped and insulated by gas hydrates." Kamata S. et al. (2019) Nature Geoscience, 12: 407–410. Further insights into Pluto's geology.

"Powering Triton's recent geological activity by obliquity tides" By F. Nimmo, J. R. Spencer. *Icarus*, 246 (2015) 2–10. A detailed insight into the obliquity tides that provide energy to Neptune's moon.

"Returning Samples from Enceladus for Life Detection" By Marc Neveu et al. (2020) Frontiers in Astronomy and Space Sciences. An indepth view on a potential mission to return samples from Enceladus.

"Salt partitioning between water and high-pressure ices. Implication for the dynamics and habitability of icy moons and water-rich planetary bodies" by Baptiste Journaux, Isabelle Daniel, Sylvain Petitgirard, Hervé Cardon, Jean-Philippe Perrillat, Razvan Caracas, and Mohamed Mezouar. *Earth and Planetary Science Letters* 463 (2017) 36–47. Assessing the effects of salts on the physical properties of high-pressure ices and therefore the possible chemical exchanges and habitability inside water-rich planetary bodies.

Searching for Life in an Ocean World: The Enceladus Life Signatures and Habitability (ELSAH) mission concept. Jennifer L Eigenbrode, Robert Gold, Christopher P. McKay, Alfonso Davila & T. Hurford. An in depth view on a potential mission to discover life on Enceladus.

"Second genesis: The search for life on other worlds" by Christopher P. McKay. *Biochemical Society*, December 2014. This article provides a nice introduction to the possibilities of life outside of our planet from a biochemistry point of view.

"Science Objectives for Flagship-Class Mission Concepts for the Search for Evidence of Life at Enceladus." (2022) Shannon MacKenzie et al. Astrobiology, Volume 22, Number 6.

"The carbonate geochemistry of Enceladus' ocean" Glein C. R., and Waite J. H. (2020) Geophysical Research Letters, 47: e2019GL085885 Provides a hypothesis on what geological processes might be present under Enceladus' ice crust.

"The Compositions of Kuiper Belt Objects" by Michael Brown. *Annual Review of Earth and Planetary Sciences,* March 2012. The author reviews the large quantity of data we have gathered on Kuiper Belt objects and suggests a framework within which we can better understand them.

"The Evolution of Icy Satellite Interiors and Surfaces" by Guy J.Consolmagno & John S.Lewis. *Icarus*, Volume 34, Issue 2, May 1978, pp. 280–293. A pivotal paper on the existence of subsurface oceans in icy satellites.

"The Possible Origin and Persistence of Life on Enceladus and Detection of Biomarkers in the Plume" by Christopher P. McKay, Carolyn C. Porco, Travis Altheide, Wanda L. Davis, and Timothy A. Kral. *ASTROBIOLOGY,* Volume 8, Number 5, 2008. A thorough review on how Cassini's instruments could have detected plausible evidence for life by analysis of hydrocarbons in the plume during close encounters.

"The Search for Life in Our Solar System and the Implications for Science and Society" by Christopher P. McKay. *Philosophical Transactions of the Royal Society,* January 2011. A summary of our efforts to search for life in our Solar System and its impact once found.

"Tidal Heating in Icy Satellite Oceans" by Chen, F. Nimmo & G.A. Glatzmaier. *Icarus,* October 2013. A thorough review of the tidal heating process in icy satellites. Don't let the math scare you; the text provides enough clarity for it to be understood within the given context.

Titan as Revealed by the Cassini Radar. Lopes, R. M. C. et al. Space Sci. Rev. 215, 33 (2019). Provides insights into one of the most remarkable planetary bodies in our Solar System.

"Vacant Habitats in the Universe" by Charles Cockell. *Trends in Ecology and Evolution,* February 2011, Vol. 26, No. 2. Overview of habitats in which geochemical processes occur without a biota, but in which the physical environmental conditions approximate to conditions in past or present terrestrial habitats.

"Vision and Voyages for Planetary Science in the Decade 2013–2022." The National Academies Press. The decadal survey that provides a strategy for the exploration of our Solar System as recommended by the U. S. scientific community.

Index